Portfolio Optimization

CRC Press
Taylor & Francis Group
6000 Broken Sound Parkway NW, Suite 300
Boca Raton, FL 33487-2742

© 2010 by Taylor & Francis Group, LLC
CRC Press is an imprint of Taylor & Francis Group, an Informa business

No claim to original U.S. Government works

Printed in the United States of America on acid-free paper
Version Date: 20110725

International Standard Book Number: 978-1-4200-8584-6 (Hardback)

Library of Congress Cataloging-in-Publication Data

Best, Michael J.
    Portfolio optimization / Michael J. Best.
        p. cm. -- (Chapman & Hall/CRC finance series)
    Includes bibliographical references and index.
    ISBN 978-1-4200-8584-6 (hardcover : alk. paper)
    1. Portfolio management. 2. Investment analysis. 3. Stocks. 4. Investments. I. Title.
II. Series.

HG4529.5.B515 2010
332.63'2042--dc22   2009053431

Visit the Taylor & Francis Web site at
http://www.taylorandfrancis.com

and the CRC Press Web site at
http://www.crcpress.com

CHAPMAN & HALL/CRC FINANCE SERIES

Series Editor

Michael K. Ong

Stuart School of Business
Illinois Institute of Technology
Chicago, Illinois, U. S. A.

Aims and Scopes

As the vast field of finance continues to rapidly expand, it becomes increasingly important to present the latest research and applications to academics, practitioners, and students in the field.

An active and timely forum for both traditional and modern developments in the financial sector, this finance series aims to promote the whole spectrum of traditional and classic disciplines in banking and money, general finance and investments (economics, econometrics, corporate finance and valuation, treasury management, and asset and liability management), mergers and acquisitions, insurance, tax and accounting, and compliance and regulatory issues. The series also captures new and modern developments in risk management (market risk, credit risk, operational risk, capital attribution, and liquidity risk), behavioral finance, trading and financial markets innovations, financial engineering, alternative investments and the hedge funds industry, and financial crisis management.

The series will consider a broad range of textbooks, reference works, and handbooks that appeal to academics, practitioners, and students. The inclusion of numerical code and concrete real-world case studies is highly encouraged.

Published Titles

Decision Options®: The Art and Science of Making Decisions, **Gill Eapen**

Emerging Markets: Performance, Analysis, and Innovation, **Greg N. Gregoriou**

Introduction to Financial Models for Management and Planning, **James R. Morris and John P. Daley**

Pension Fund Risk Management: Financial and Actuarial Modeling, **Marco Micocci, Greg N. Gregoriou, and Giovanni Batista Masala**

Stock Market Volatility, **Greg N. Gregoriou**

Portfolio Optimization, **Michael J. Best**

Proposals for the series should be submitted to the series editor above or directly to:
CRC Press, Taylor & Francis Group
4th, Floor, Albert House
1-4 Singer Street
London EC2A 4BQ
UK

CHAPMAN & HALL/CRC FINANCE SERIES

Portfolio Optimization

CV 08.12.2019 1753

Michael J. Best
University of Waterloo
Ontario, Canada

CRC Press
Taylor & Francis Group
Boca Raton   London   New York

CRC Press is an imprint of the
Taylor & Francis Group, an **informa** business

A CHAPMAN & HALL BOOK

Chapman & Hall/CRC
Taylor & Francis Group
6000 Broken Sound Parkway NW, Suite 300
Boca Raton, FL 33487-2742

© 2010 by Taylor and Francis Group, LLC
Chapman & Hall/CRC is an imprint of Taylor & Francis Group, an Informa business

No claim to original U.S. Government works

International Standard Book Number: 978-1-4200-8584-6 (Hardback)

Library of Congress Cataloging-in-Publication Data

Best, Michael J.
    Portfolio optimization / Michael J. Best.
        p. cm. -- (Chapman & Hall/CRC finance series)
    Includes bibliographical references and index.
    ISBN 978-1-4200-8584-6 (hardcover : alk. paper)
    1. Portfolio management. 2. Investment analysis. 3. Stocks. 4. Investments. I. Title.
II. Series.

    HG4529.5.B515 2010
    332.63'2042--dc22   2009053431

Visit the Taylor & Francis Web site at
http://www.taylorandfrancis.com

and the CRC Press Web site at
http://www.crcpress.com

To

my wife *Patti*

and

my daughters *Jiameng, Zhengzheng, Jing, Xianzhi, and Thamayanthi*

Contents

Preface

This book is designed as a first course in Markowitz Mean-Variance portfolio optimization. The sole prerequisite is elementary linear algebra: linear independence, matrix vector operations and inverse matrices. The presentation is quantitative and all material is developed precisely and concisely. Part of the author's motivation in writing this book was his realization that the mathematical tools of linear algebra and optimization can be used to formulate and present the key ideas of the subject very quickly and cleanly. A unique feature of the book is that it extends the key ideas of the Markowitz "budget constraint only" model to a model having linear inequality constraints. The linearly constrained model is what practitioners insist upon using.

The author is a researcher in, and a practitioner of, portfolio optimization. This book reflects that background.

Chapter 1 presents necessary and sufficient conditions for optimality for quadratic minimization subject to linear equality constraints. These results are a corner stone for the book. Optimality conditions (Karush-Kuhn-Tucker conditions) for linearly constrained nonlinear problems are also given as they are required in the development of linearly constrained Sharpe ratios and implied risk free rates. Extreme points are also developed as they play a key role in certain aspects of portfolio optimization.

Chapter 2 develops the key properties of the efficient frontier by posing the portfolio optimization problem as a parametric quadratic minimization problem and then solving it explicitly using results from Chapter 1. The equation of the efficient frontier follows quickly from this. Alternate derivations (maximization of expected return and minimization of risk) are formulated as exercises. Chapter 3 extends these results to problems having a risk free asset. This gives the Capital and Security Market lines, again by using the optimality conditions of Chapter 1. The tangency result for the Capital Market Line and the efficient frontier for risky assets is shown in quick and precise detail.

Chapter 4 develops Sharpe ratios and implied risk free rates in two ways. The first is by direct construction. The tangent to the efficient frontier is formulated and the implied risk free rate is calculated. The argument is reversible so that if the risk free rate is specified, the market portfolio can then be calculated. The second approach formulates each of these two problems as nonlinear optimization problems and the nonlinear optimization techniques of Chapter 1 are used to solve them. The advantage of this approach is that it is easy to generalize to problems having linear constraints as is done in Chapter 9. Finally, it is shown that any point on the efficient frontier gives an optimal solution to four portfolio optimization problems involving portfolio expected return, variance, risk free rates and market portfolios.

A knowledge of quadratic programming is essential for those who wish to solve practical portfolio optimization problems. Chapter 5 introduces the key concepts in a geometric way. The optimality conditions (Karush-Kuhn-Tucker conditions) are first formulated from graphical examples, then formulated algebraically and finally they are proven to be sufficient for optimality. A simple quadratic programming algorithm is formulated in Chapter 6. Its simplicity stems from leaving the methodology of solving the relevant linear equations unspecified and just focuses on determining the search direction and step size. The algorithm is implemented in a MATLAB® program called QPSolver and the details of the method are given at the end of the chapter.

Chapter 7 begins with an example of a problem with no short sales constraints and solves it step by step for all possible values of the risk aversion parameter. It is observed that the efficient portfolios are piecewise linear functions of the risk aversion parameter. This result is then shown to be true in general for a constrained portfolio optimization problem. The resulting efficient frontier is shown to be piecewise parabolic and (normally) differentiable. Chapter 8 extends these results into an algorithm which determines the entire (constrained) efficient frontier, its corner portfolios, the piecewise linear expected returns and the piecewise quadratic variances. This then is formulated as a MATLAB program and is explained in detail at the end of Chapter 8.

Chapter 9 extends Sharpe ratios and implied risk free returns to portfolio optimization problems which have linear inequality constraints and shows how to determine them. Some of the results are perhaps surprising. For example, when a corner portfolio happens to be an extreme point (a simple example is given), the efficient frontier is not differentiable at this point. Further, if this portfolio is taken to be the market portfolio, the implied risk free return is no longer a unique point but is actually an infinite range of values (formulae are given).

The text shows clearly how to implement each technique by hand (for learning purposes). However, it is anticipated that the reader will want to apply these techniques to problems of their own choosing when solving by hand is prohibitive. Thus each chapter concludes with the presentation of several MATLAB programs designed to implement the methods of the chapter. These programs are included on a CD which accompanies the book. Each program is explained, line by line, in the book itself. It is anticipated that a reader not familiar with MATLAB (nor even computer programming) can easily use these programs to good advantage. One such program, upon being given the problem covariance matrix and expected return vector, computes the coefficients of the efficient frontier. For a portfolio optimization with arbitrary linear constraints, a second program computes any point on the efficient frontier for a specified expected risk aversion parameter. A third program constructs the entire efficient frontier (corner points) for a linearly constrained problem. The book includes numerous exercises. Some of these are numerical and are designed to amplify concepts in the text with larger numerical examples. Others are more analytical and will exercise and enhance the reader's understanding of the textual material. The prerequisite information needed for some exercises is set up in the text and the exercises take it to its logical conclusion.

Acknowledgments

A book like this could not have been written without the help of many people. Special thanks go to Professor Robert R. Grauer, who introduced me to the subject of Portfolio Optimization many years ago and taught me the basics from a qualitative point of view. In addition, many people generously contributed their time, thought and intellectual energy to this book. These wonderful people include Patricia M. Best, Thamayanthi Chellathurai, Yuen-Lam Cheung, Duncan Hamre, Jaraslova Hlouskova, Darryl Shen, Jing Wang, Xianxhi Wang and Xili Zhang.

About the Author

Michael J. Best received his Ph.D. from the Department of Industrial Engineering and Operations Research at the University of California, Berkeley in 1971. Since then, he has been with the Department of Combinatorics and Optimization, University of Waterloo, Waterloo, Ontario, Canada. He has written over 37 papers on finance and nonlinear programming and coauthored a textbook on linear programming. During this time, he has also been a consultant to many financial institutions in the area of large scale portfolio optimization algorithms and software. These institutions include Bank of America, Ibbotson Associates, Montgomery Assets Management, Deutsche Bank, Toronto Dominion Bank and Black Rock-Merrill Lynch.

Chapter 1

Optimization

1.1 Quadratic Minimization

Throughout this chapter, we will deal with the minimization of a quadratic function subject to linear equality constraints. In Chapters 2−4, this same type of optimization problem plays a central role. It is therefore essential to have available necessary and sufficient conditions for a point to minimize a quadratic function subject to linear equality constraints. It is the purpose of this section to develop such conditions.

Throughout this text, we will rely heavily on vector and matrix notation. Prime ($'$) will be used to denote matrix transposition. All unprimed vectors will be column vectors. The i−th component of the vector x_0 will be denoted by $(x_0)_i$. In a numerical example, we sometimes use the symbol times ($\times$) to indicate multiplication. In 2-dimensional examples, we generally use the problem variables x_1 and x_2 with $x = (x_1, x_2)'$. In other situations, x_1 and x_2 may denote two n−vectors. The meaning should be clear from the context. The end of a proof is signified by an unfilled box ($\square$). The end of an example is indicated by an unfilled diamond ($\Diamond$).

First, we need the concept of the gradient of a function. A general quadratic function of n variables can be written

$$f(x) = \sum_{i=1}^{n} c_i x_i + \frac{1}{2} \sum_{i=1}^{n} \sum_{j=1}^{n} \gamma_{ij} x_i x_j.$$

Letting $c = (c_1, c_2, \ldots, c_n)'$, $x = (x_1, x_2, \ldots, x_n)'$ and $C = [\gamma_{ij}]$, $f(x)$ can be written more compactly as

$$f(x) = c'x + \frac{1}{2} x'Cx. \tag{1.1}$$

We assume that C is symmetric (for if it is not, it may be replaced by $\frac{1}{2}[C + C']$ which is symmetric and leaves the value of f unchanged.) The gradient of f at x is the n-vector of first partial derivatives of f evaluated at x and is denoted by $\nabla f(x)$. It is not hard to show (Exercise 1.4) that for the quadratic function (1.1),

$$\nabla f(x) = c + Cx. \tag{1.2}$$

In the material to follow, C will always be a covariance matrix and as such, is symmetric and positive semidefinite (C *positive semidefinite* $\Leftrightarrow s'Cs \geq 0$ for all s). Sometimes we will require that C be positive definite (C *positive definite* $\Leftrightarrow s'Cs > 0$ for all $s \neq 0$.) Throughout these sections we will assume that C is positive semidefinite.

Let us look at an example.

Example 1.1
Give a graphical solution to

$$\min\{ -4x_1 - 4x_2 + x_1{}^2 + x_2{}^2 \mid x_1 + x_2 = 2\}.$$

Here,

$$c = \begin{bmatrix} -4 \\ -4 \end{bmatrix} \text{ and } C = \begin{bmatrix} 2 & 0 \\ 0 & 2 \end{bmatrix}.$$

The objective function for this problem (the function to be minimized) can be written as $f(x) = (x_1 - 2)^2 + (x_2 - 2)^2 - 8$ so that its level sets

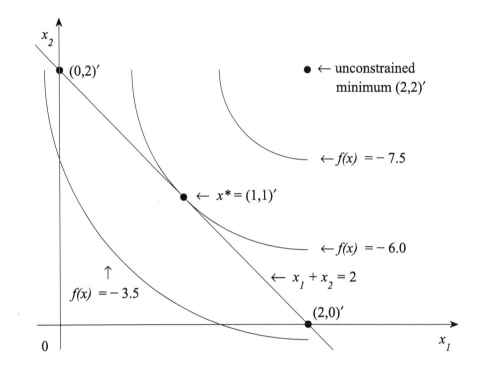

Figure 1.1   Geometry of Example 1.1.

(the points where the function has constant value) are circles centered at $(2,2)'$. Figure 1.1 illustrates the geometry of this example. Points on the line $x_1 + x_2 = 2$ between $(2,0)'$ and $(0,2)'$ all give smaller objective function values than -3.5, so we can find points which reduce f by "shrinking" the level sets further. However, if they are shrunk too much they will not intersect the line $x_1 + x_2 = 2$. See the level set $f(x) = -7.5$ in Figure 1.1. The optimal solution is obtained when the level set is shrunk until it intersects the line $x_1 + x_2 = 2$ at just a single point. From Figure 1.1, this occurs at $(1,1)'$ for the level set $f(x) = -6$ and the optimal solution is $x^* = (1,1)'$.

We can use Example 1.1 to deduce an important characterization of an optimal solution in terms of the gradient of the objective function and the gradient of the constraint function. Recall that the gradient of

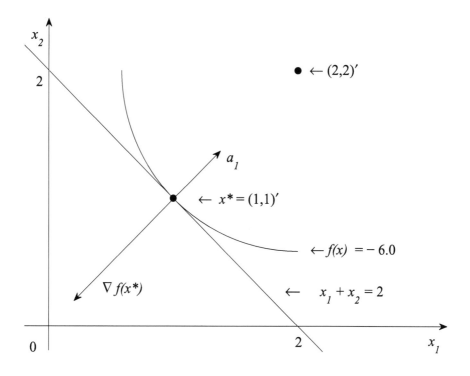

Figure 1.2　Optimality Conditions for Example 1.1.

a function at a point, points in the direction of maximum local increase for the function and is orthogonal to the tangent of the function at the point. $\nabla f(x^*)$ is shown in Figure 1.2. Let a_1 denote the gradient of the constraint function $x_1 + x_2 = 2$. Then $a_1 = (1,1)'$ and is also shown in Figure 1.2. Now we make the critical observation. At the optimal solution, $\nabla f(x^*)$ is a multiple of a_1.

Indeed, this condition together with the problem constraint completely defines the optimal solution. Suppose we did not know the optimal solution for Example 1.1. Consider the two conditions

$$x_1 + x_2 = 2, \quad \text{and} \quad \nabla f(x) = u_1 a_1, \tag{1.3}$$

where the second condition requires the gradient of f to be a scalar multiple (namely u_1) of a_1. Rearrangement of the terms in (1.3) gives

$$
\begin{array}{rcrcl}
2x_1 & & & - & u_1 & = & 4, \\
& & 2x_2 & - & u_1 & = & 4, \\
x_1 & + & x_2 & & & = & 2.
\end{array}
$$

The solution of this is $x^* = (1,1)'$ and $u_1 = -2$. This shows that the optimal solution is part of the solution to three linear equations in three unknowns.   ◇

Our next goal is to generalize these concepts to problems having n variables and m constraints. But first we need to formulate Taylor's Series for quadratic functions of n variables.

Theorem 1.1 *(Taylor's Theorem) Let $f(x) = c'x + \frac{1}{2}x'Cx$. Then for any points x and x_0,*

$$
f(x) = f(x_0) + \nabla f(x_0)'(x - x_0) + \frac{1}{2}(x - x_0)'C(x - x_0).
$$

There are many forms of Taylor's Theorem. The Taylor's series for quadratic functions is especially simple as there is no remainder term. It is easy to prove Theorem 1.1 by simply multiplying out the terms on the right and simplifying (see Exercise 1.5).

Consider the model problem

$$
\min\{c'x + \frac{1}{2}x'Cx \mid Ax = b\}, \tag{1.4}
$$

where c and x are $(n,1)$, C is (n,n), symmetric and positive semidefinite, A is (m,n) and b is $(m,1)$. Thus our model problem has n variables and m equality constraints. The constraints can equally well be written as

$$
a_i'x = b_i, \quad i = 1, 2, \ldots, m,
$$

where each a_i is an n-vector,

$$
A' = [a_1, a_2 \ldots, a_m] \quad \text{and} \quad b = (b_1, b_2, \ldots, b_m)'.
$$

In this format, a_i is the gradient of the $i-th$ constraint function.

In Example 1.1, the key condition was that $\nabla f(x)$ be a multiple of the gradient of the constraint. The generalization of this for (1.4) is that

$$\nabla f(x) \; = \; u_1 a_1 \; + \; u_2 a_2 \; + \; \cdots \; + \; u_m a_m,$$

which says that $\nabla f(x)$ must be a linear combination of the gradients of the constraints. By replacing u_i with $-u_i$, this condition becomes

$$- \nabla f(x) \; = \; u_1 a_1 \; + \; u_2 a_2 \; + \; \cdots \; + \; u_m a_m. \qquad (1.5)$$

Later, when we deal with optimality conditions for problems with inequality constraints it will be natural to formulate them in terms of $-\nabla f(x)$ rather than $\nabla f(x)$. To be compatible with those conditions, we will formulate the optimality conditions for (1.4) in terms of $-\nabla f(x)$. Note that with matrix notation, (1.5) can be written quite compactly:

$$-\nabla f(x) \; = \; A'u,$$

where $u = (u_1, u_2, \ldots, u_m)'$.

Definition 1.1 *The* **optimality conditions** *for (1.4) are (1) $Ax_0 = b$ and (2) there exists a vector u with $-\nabla f(x_0) = A'u$.*

The vector u in Definition 1.1 is called the *multiplier vector* for the problem. It has one component for each row of A.

The optimality conditions for (1.4) are a special case of the Karush-Kuhn-Tucker conditions for more general nonlinear programming problems. There is one component of u for each constraint. The number u_i is called the *multiplier* associated with constraint i. The first optimality condition, $Ax_0 = b$ is usually called *primal feasibility* and the second optimality condition, $-\nabla f(x_0) = A'u$ is generally called *dual feasibility*.

The key result here is that the optimality conditions for (1.4) are both necessary and sufficient for optimality.

Theorem 1.2 x_0 *is optimal for (1.4) if and only if* x_0 *satisfies the optimality conditions for (1.4).*

Proof: We will prove only the sufficiency part. Necessity is a bit more complicated (see Mangasarian [18] for details).

Assume then that x_0 satisfies the optimality conditions for (1.4). To show that x_0 is optimal, we must show that $f(x_0) \leq f(x)$ for all x satisfying $Ax = b$. Let x be such that $Ax = b$. By Taylor's Theorem,

$$f(x) = f(x_0) + \nabla f(x_0)'(x - x_0) + \frac{1}{2}(x - x_0)'C(x - x_0). \quad (1.6)$$

From the optimality conditions, $\nabla f(x_0) = -A'u$ so that

$$\nabla f(x_0)'(x - x_0) = -u'A(x - x_0) = -u'(b - b) = 0. \quad (1.7)$$

Furthermore,

$$(x - x_0)'C(x - x_0) \geq 0, \quad (1.8)$$

because C is positive semidefinite. Using (1.7) and (1.8) in (1.6) shows $f(x) \geq f(x_0)$, which is the desired result. $\qquad\square$

Theorem 1.2 can be reformulated in the following way, which shows that the optimal solution for (1.4) is simply the solution of a system of linear equations.

Theorem 1.3 x_0 *is optimal for (1.4) if and only if there exists an* $m-$*vector* u *such that* (x_0, u) *satisfies the linear equations*

$$\begin{bmatrix} C & A' \\ A & 0 \end{bmatrix} \begin{bmatrix} x_0 \\ u \end{bmatrix} = \begin{bmatrix} -c \\ b \end{bmatrix}.$$

Proof: Multiplying out the two partitions of the equations in the statement of the theorem, the result follows directly from Theorem 1.2. $\qquad\square$

Note that the coefficient matrix of the linear equations in Theorem 1.3 is square and has dimension $(n + m, n + m)$.

We complete this section with an example illustrating Theorem 1.3.

Example 1.2
Use Theorem 1.3 to solve (1.4) with data

$$c = (1.1, 1.2, 1.3)', \quad C = diag(1 \times 10^{-2}, 5 \times 10^{-2}, 7 \times 10^{-2}),$$

and

$$A = \begin{bmatrix} 1 & 1 & 1 \\ 0.5 & 0.5 & 0.0 \end{bmatrix}, \quad b = (1, 1)'.$$

According to Theorem 1.3, x_0 will be optimal for this problem if and only if x_0 together with a 2−vector u satisfies the linear equations

$$\begin{bmatrix} 1 \times 10^{-2} & 0 & 0 & 1 & 0.5 \\ 0 & 5 \times 10^{-2} & 0 & 1 & 0.5 \\ 0 & 0 & 7 \times 10^{-2} & 1 & 0 \\ 1 & 1 & 1 & 0 & 0 \\ 0.5 & 0.5 & 0 & 0 & 0 \end{bmatrix} \begin{bmatrix} x_0 \\ u \end{bmatrix} = \begin{bmatrix} -1.1 \\ -1.2 \\ -1.3 \\ 1 \\ 1 \end{bmatrix}. (1.9)$$

A MATLAB program will be given in Section 1.4 which will numerically solve this problem. ◊

1.2   Nonlinear Optimization

In Section 1.1, we developed optimality conditions for minimizing a quadratic function of n variables subject to a number of linear equality constraints. In order to obtain the results in Chapter 4, we need to be able to characterize an optimal solution for minimizing a nonlinear function of n variables subject to linear equality constraints, and that is precisely what we do in this section. Our presentation here generalizes that of Section 1.1 from quadratic functions to more general nonlinear functions.

First we need the notion of a Hessian matrix. Let $f(x)$ be a function of the n−vector x having continuous second partial derivatives. Then the *Hessian matrix of f at x* is the (n, n) matrix

$$H(x) = \left[\frac{\partial^2 f(x)}{\partial x_i \partial x_j} \right];$$

i.e.; $H(x)$ is the matrix whose (i, j)th element is the second partial derivative of f with respect to x_j and then x_i. Note that if f is the quadratic function $f(x) = c'x + \frac{1}{2}x'Cx$, $H(x) = C$ and is independent of x.

Theorem 1.4 *(Taylor's Theorem) Let $f(x)$ have continuous second partial derivatives. Then for any points x and x_0,*

$$f(x) = f(x_0) + \nabla f(x_0)'(x - x_0) + \frac{1}{2}(x - x_0)'H(\xi)(x - x_0)$$

where $\xi = \theta x_0 + (1 - \theta)x$, for some scalar θ with $0 \le \theta \le 1$.

Proof: Define the function of a single variable σ as

$$h(\sigma) = f(x_0 + \sigma(x - x_0)).$$

Applying Taylor's Theorem for a function of a single variable, between the points 1 and 0 gives

$$h(1) = h(0) + \frac{d}{d\sigma}h(0)(1 - 0) + \frac{1}{2}\frac{d^2}{d\sigma^2}h(\theta)(1 - 0)^2, \qquad (1.10)$$

where $0 \le \theta \le 1$. Applying the chain rule to the derivatives in (1.10) gives

$$h(1) = h(0) + \sum_{i=1}^{n} \frac{\partial f(x_0)}{\partial x_i}(x - x_0)_i + \frac{1}{2}\sum_{i=1}^{n}\sum_{j=1}^{n} \frac{\partial^2 f(\xi)}{\partial x_j \partial x_i}(x - x_0)_i(x - x_0)_j,$$

where $\xi = \theta x + (1 - \theta)x_0$. Observing that $h(1) = f(x)$, $h(0) = f(x_0)$, expressing the above summations in vector notation and noting that $0 \le \theta \le 1$ implies $0 \le (1 - \theta) \le 1$, the theorem follows from this last equation. □

We next turn to the question of characterizing an optimal solution for the model problem

$$\min\{f(x) \mid Ax = b\,\}, \tag{1.11}$$

where x is an $n-$vector, $f(x)$ is a function of x, A is an (m,n) matrix and b is an m-vector. Based on Definition 1.1 for quadratic functions, one might expect the optimality conditions for (1.11) to be (1) $Ax_0 = b$, and (2) there exists an $m-$vector u with $-\nabla f(x_0) = A'u$. For sufficiency of these conditions, we also require that $H(x_0)$ be positive definite as shown in the following theorem.

Theorem 1.5 *Suppose (1) $Ax_0 = b$, (2) there exists an $m-$vector u with $-\nabla f(x_0) = A'u$, and (3) $H(x_0)$ is positive definite, then x_0 is a local minimizer for (1.11). In addition, if $H(x)$ is positive definite for all x such that $Ax = b$, then x_0 is a global minimizer for (1.11).*

Proof: Let x be any point such that $Ax = b$. Then from (1) and (2) in the statement of the theorem,

$$
\begin{aligned}
-\nabla f(x_0)'(x - x_0) &= (A'u)'(x - x_0) \\
&= u'A(x - x_0) \\
&= u'(Ax - Ax_0) \\
&= u'(b - b) \\
&= 0.
\end{aligned}
$$

Thus,

$$-\nabla f(x_0)'(x - x_0) = 0. \tag{1.12}$$

Now using Taylor's Theorem (Theorem 1.4), we have

$$f(x) = f(x_0) + \nabla f(x_0)'(x - x_0) + \frac{1}{2}(x - x_0)'H(\xi)(x - x_0), \tag{1.13}$$

where $\xi = \theta x_0 + (1 - \theta)x$, for some scalar θ with $0 \le \theta \le 1$. Using (1.12) in (1.13) gives

$$f(x) = f(x_0) + \frac{1}{2}(x - x_0)'H(\xi)(x - x_0). \tag{1.14}$$

If $H(x_0)$ is positive definite, then $H(x)$ is also positive definite in some small neighborhood of x_0 and with (1.14) this implies $f(x_0) \le f(x)$ for all x with $Ax = b$ in a small neighborhood of x_0; i.e., x_0 is a local minimizer. In addition, if $H(x)$ is positive definite for all x with $Ax = b$, (1.14) implies that $f(x_0) \le f(x)$ for all x with $Ax = b$ and in this case x_0 is a global minimizer. □

The reader may recall the elementary Calculus formulae for the product and ratio of two functions:

$$d(uv) = vdu + udv \text{ and } d\left[\frac{u}{v}\right] = \frac{vdu - udv}{v^2}.$$

We next formulate generalizations of these two rules for functions of n variables.

Theorem 1.6 *Let $f(x)$ and $g(x)$ be continuously differentiable functions of the $n-$vector x. Then*

(a) $\nabla(f(x)g(x)) = f(x)\nabla g(x) + g(x)\nabla f(x)$,

(b) $\nabla\left[\frac{f(x)}{g(x)}\right] = \left[\frac{g(x)\nabla f(x) - f(x)\nabla g(x)}{g^2(x)}\right]$, *provided $g \ne 0$ in some neighborhood of x.*

Proof: For Part (a), let $h(x) = f(x)g(x)$. Then using the product rule for functions of a single variable and for $i = 1, 2, \ldots, n$,

$$\frac{\partial h(x)}{\partial x_i} = f(x)\frac{\partial g(x)}{\partial x_i} + g(x)\frac{\partial f(x)}{\partial x_i}.$$

Expressing this last in vector notation gives the desired result.

For Part (b), let

$$h(x) = \frac{f(x)}{g(x)}.$$

Then using the quotient rule for functions of a single variable and for $i = 1, 2, \ldots, n$,

$$\frac{\partial h(x)}{\partial x_i} = \frac{g(x)\frac{\partial f(x)}{\partial x_i} - f(x)\frac{\partial g(x)}{\partial x_i}}{g^2(x)}.$$

Expressing this last in vector notation gives the desired result. □

1.3   Extreme Points

Extreme points usually arise in the context of a Linear Programming (LP) problem. An LP is the problem of minimizing a linear function subject to linear inequality/equality constraints. Extreme points also play an important role in portfolio optimization and we formulate them here. First we introduce the basic ideas by means of an example.

Example 1.3
  Sketch the region defined by the linear inequalities

$$\left.\begin{array}{rrrrr} x_1 & + & 3x_2 & \leq & 9, \quad (1) \\ 2x_1 & + & x_2 & \leq & 8, \quad (2) \\ - & x_1 & & \leq & -1, \quad (3) \\ & - & x_2 & \leq & -1 \quad (4) \end{array}\right\}. \tag{1.15}$$

The set of points which satisfies constraint (1) is the set of points which is below and to the left of the line $x_1 + 3x_2 = 9$ while those that satisfy constraint (2) are those below and to the left of the line $2x_1 + x_2 = 8$. The set of points which satisfies (3) and (4) is those which lie above the line $-x_2 = -1$ and those which lie to the right of the line $-x_1 = -1$, respectively. The set of all points which simultaneously satisfies (1) to (4) is called the feasible region for (1.15) and is shown as the shaded region in Figure 1.3.

  The *extreme points* (or *corner points*) P_1, P_2, P_3 and P_4 in Figure 1.3 play a special role. For example, it is clear that the minimum value of $-x_2$ (i.e., the maximum value of x_2) occurs at P_1 and the minimum value of $-x_1$ (i.e., the maximum value of x_1) occurs at P_3. A little thought shows that the minimum value of the linear function $-x_1 - x_2$ occurs at P_2.

  Notice in this **2**-dimensional example that each extreme point is the solution of **2** of the inequalities of (1.15) holding as equalities. For example, P_1 is the solution of a system of *two* simultaneous linear equations

$$\begin{array}{rrrrr} x_1 & + & 3x_2 & = & 9, \\ - & x_1 & & = & -1, \end{array}$$

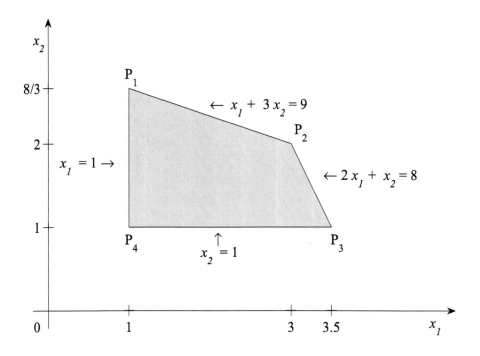

Figure 1.3   Feasible Region for Example 1.3.

and P_2 is the solution of

$$\begin{aligned} x_1 &+ 3x_2 &= 9, \\ 2x_1 &+ x_2 &= 8. \end{aligned}$$

Thus $P_1 = (1, 8/3)'$ and $P_2 = (3, 2)'$.   ◇

We next consider extreme points for a problem with n variables and m constraints. In particular, consider the feasible region R defined by

$$R = \{\, x \mid a_i'x \le b_i,\ i = 1, 2, \ldots, m \,\},$$

where each a_i is a given n–vector and each b_i is a given scalar, $i = 1, 2, \ldots, m$. R can be written more compactly as

$$R = \{\, x \mid Ax \le b \,\},$$

where $A = [a_1, a_2, \ldots, a_m]'$ and $b = (b_1, b_2, \ldots, b_m)'$. The vector a_i is the gradient of the i-th constraint function for $i = 1, 2, \ldots, m$.

Definition 1.2 *A point x_0 is* **feasible** *for R if $x_0 \in R$ and* **infeasible** *otherwise.*

Definition 1.3 *For any i with $1 \leq i \leq m$, constraint i is* **active** *at x_0 if $a_i' x_0 = b_i$ and* **inactive** *at x_0 if $a_i' x_0 < b_i$*

By analogy to the 2-dimensional case, we make the following definition.

Definition 1.4 *A point x_0 is an* **extreme point** *of R if $x_0 \in R$ and there are n constraints having linearly independent gradients active at x_0.*

The linear independence requirement for an extreme point as well as degenerate extreme points is developed in Exercises 1.12 and 1.13.

Not all feasible regions possess extreme points. The following result gives necessary and sufficient conditions for the existence of an extreme point (see Exercise 2.18 of Best and Ritter [17]).

Theorem 1.7 *Let A be an (m, n) matrix and suppose that the set $R = \{x \mid Ax \leq b\}$ is nonempty. Then R has extreme points if and only if rank $(A) = n$.*

We next briefly consider the LP

$$\min\{ c'x \mid Ax \leq b \}, \tag{1.16}$$

which finds the minimum value of the linear function $c'x$ over all points in R and c is a given vector. A key result from linear programming is the following (see Exercise 3.13 Best and Ritter [17]).

Theorem 1.8 *For (1.16), assume that rank $(A) = n$ and that the problem is bounded from below. Then one of the extreme points of (1.16) is an optimal solution for it.*

Theorem 1.8 says that an LP may be solved by just looking at its extreme points. We illustrate this in the following example.

Example 1.4
Solve
$$\min\{-\mu'x \mid x \geq 0,\ l'x = 1\},$$
where $l = (1, 1, \ldots, 1)'$, μ_i is the expected return on asset i and x_i is the proportion of wealth invested in asset i, for $i = 1, 2, \ldots, n$, $\mu = (\mu_1, \mu_2, \ldots, \mu_n)'$ and $x = (x_1, x_2, \ldots, x_n)'$.

Minimizing $-\mu'x$ is equivalent to maximizing $\mu'x$ so the given problem is equivalent to maximizing the total expected return $\mu'x = \mu_1 x_1 + \mu_2 x_2 + \ldots + \mu_n x_n$ subject to "no short sales" restrictions ($x \geq 0$) and the budget constraint which requires the sum of the proportions to be 1.

Observe that the nonnegativity constraints $x \geq 0$ are just a compact way of writing $x_i \geq 0$, $i = 1, 2, \ldots, n$. Thus, this problem has $n + 1$ constraints and each such extreme point must have n active constraints. The budget constraint is always active and so from Definition 1.4, $n - 1$ of the nonnegativity constraints must be active at any extreme point. Thus the extreme points for this problem are the unit vectors $(1, 0, \ldots, 0)'$, $(0, 1, 0, \ldots, 0)'$,$\ldots$, $(0, 0, \ldots, 0, 1)'$. Theorem 1.8 states the optimal solution will be that extreme point which gives the smallest objective function value. The objective function values for these extreme points are just $-\mu_1$, $-\mu_2$, $\ldots$, $-\mu_n$ and the smallest of these occurs for that index k such that $\mu_k = \max\{\mu_1, \mu_2, \ldots, \mu_n\}$ and the optimal holdings is x^* where $(x^*)_i = 0$, $i = 1, \ldots, n$, $i \neq k$ and $(x^*)_k = 1$.

Another way of saying this is that if one is trying to maximize his/her expected return subject to no short sales constraints, the optimal solution is to invest all wealth into the asset with highest expected return. $\diamond$

1.4   Computer Results

At the end of each chapter of this text, we provide MATLAB programs to perform the calculations of the methods developed in the chapter. To obtain a deeper understanding of the various methods presented, it is important to be able to solve small problems by hand. However, to appreciate the financial implications of the results obtained, it is important to apply the theory to differing data sets and quickly determine the results. The MATLAB codes provided at the end of each chapter provide this capability. The sample data used to illustrate each MATLAB program can easily be replaced by whatever data the reader wishes.

Each MATLAB program consists of one or more m−files. These are text files whose name has the extension "m." The m−files used in this text are contained in the accompanying CD.

We illustrate the process by formulating a MATLAB program to solve (1.4) using Theorem 1.3 and data of Example 1.2.

The MATLAB program named "Example1p2.m" and is shown in Figure 1.4. Line 1 contains the name of the file. The "%" at the start of line 1 tells MATLAB that what follows on the line is a comment and not part of the program. Lines 2−7 define the problem data. Note that the ";" in the definition of A in line 6 signifies the beginning of a new row in the matrix.

Line 8 invokes the function "checkdata" (see Figure 1.5). This checks that "BigC" is indeed symmetric and positive definite. Line 9 constructs the partitioned coefficient matrix of Theorem 1.3 as follows. The code fragment [BigC, A'] produces the matrix $[C, A']$ and the MATLAB function "vertcat," concatenates (vertically) this matrix with the matrix $[A, 0]$, where 0 is an (m, m) matrix of zeros. The resulting matrix is thus

$$H = \begin{bmatrix} C & A' \\ A & 0 \end{bmatrix}.$$

Similarly, in line 10, the right-hand side of the linear system of Theorem 1.3, namely $\begin{bmatrix} -c \\ b \end{bmatrix}$, is constructed using vertical concatenation and stored in "rhs." The linear system is then solved in line 11 with the composite solution being stored in the $(n + m)−$ vector y. Line 12 obtains the optimal solution x as the first n components of y, and line 13 obtains the multiplier vector u from the last m components of y.

Running this program with the indicated data gives the optimal solution $x_0 = (3.3333, -1.3333, -1.0)'$ with multiplier vector $u = (-1.23, 0.1933)'$.

```
1  %Example1p2.m
2  n = 3
3  m = 2
4  c = [ 1.1 1.2 1.3 ]'
5  BigC = [1.e-2 0 0 ; 0 5.0e-2 0 ; 0 0 7.0e-2 ]
6  A = [1 1 1 ; 0.5 0.5 0 ]
7  b = [1 1 ]'
8  checkdata(BigC,1.e-6);
9  H = vertcat([BigC,A'] , [A,zeros(m,m)])
10 rhs = vertcat(-c,b)
11 y = H^-1 * rhs               % Solve Hy = rhs.
12 x = y(1:n)                   % x = optimal solution
13 u = y(n+1:n+m)               % u = multipliers for Ax = b
```

Figure 1.4   Solution of (1.4) Using Theorem 1.3: Example1p2.m.

It is not uncommon for a user to have errors in the input data. This can produce erroneous output and the user has no idea whether the problem is with the MATLAB program or the data used. The author is of the opinion that the computer program should check to see if the supplied data is reasonable. The function "checkdata.m" does this and is shown in Figure 1.5. It checks whether the input matrix "Sigma" is symmetric and positive definite as follows. Line 4 calculates the norm of (Sigma – Sigma'). If this is $\leq$ tol, the test is passed and execution passes to line 10. If this is $\geq$ tol, the error messages in lines 5 and 6 are printed and the "pause" statement in line 7 causes execution of the program to stop until "Ctrl+c" is pressed in the MATLAB Command Window.

A well known theorem in linear algebra (see for example Noble, [20]) states that a symmetric matrix is positive definite if and only if its eigenvalues are all strictly positive and such a matrix is positive semidefinite if and only if its eigenvalues are all nonnegative. Checkdata next uses this result. The code fragment "min(eig(Sigma))" computes the smallest eigenvalue of Sigma. If this is $\geq$ tol, the test is passed and execution returns to the calling program. If this is $\leq$ tol, the error message in lines 11 and 12 are printed and the program is paused as in the first test.

Most of the calculations performed in subsequent chapters will require the input of a positive definite matrix. We will always use checkdata to check the appropriate condition.

```
1   %checkdata.m
2   function checkdata(Sigma,tol)
3   format compact
4   if norm(Sigma - Sigma') > tol
5       errmsg = 'Error: covariance matrix is not symmetric'
6       soln = 'Press Ctrl+C to continue'
7       pause
8       return
9   end
10  if min(eig(Sigma))< tol
11      errmsg = 'Error: covariance matrix is not positive
12      definite'
13      soln = 'Press Ctrl+C to continue'
14      pause
15      return
16  end
```

Figure 1.5   Check for Symmetry and Positive Definiteness: checkdata.m.

1.5   Exercises

1.1 Consider the problem

$$\text{minimize}: \quad x_1^2 + 2x_2^2 + x_3^2$$
$$\text{subject to}: \quad x_1 + 2x_2 + x_3 = 4.$$

Use the optimality conditions, as in Example 1.1, to solve the problem (no graph required).

1.2 Repeat Question 1.1 for the problem

$$\begin{aligned} \text{minimize}: \quad & 2x_1^2 + 3x_2^2 + 4x_3^2 \\ \text{subject to}: \quad & 2x_1 + 3x_2 + 4x_3 = 9. \end{aligned}$$

1.3 Repeat Question 1.1 for the problem

$$\begin{aligned} \text{minimize}: \quad & x_1 + x_1^2 + 2x_2 + x_2^2 + x_3 + 2x_3^2 \\ \text{subject to}: \quad & 3x_1 + 4x_2 + 5x_3 = 12. \end{aligned}$$

1.4 Verify (equation 1.2).

1.5 Prove Theorem 1.1 (Taylor's Theorem).

1.6 The *Lagrangian* function for (1.4) is $L(x, u) = f(x) + u'(Ax - b)$. Show that setting the gradient of $L(x, u)$ with respect to u to zero gives optimality condition (1) in Definition 1.1. Furthermore, show that setting the gradient of $L(x, u)$ with respect to x gives optimality condition (2) of Definition 1.1.

1.7 Let C be an (n, n) positive definite symmetric matrix. Show, using the definition of a positive definite matrix, that

    (a) C is nonsingular.

    (b) C^{-1} is symmetric.

    (c) C^{-1} is positive definite.

Hint: A concise argument need only require a few lines.

1.8 Let H be any (n, k) matrix. Show that $C = HH'$ is positive semidefinite and that if $k = n$ and H has full row rank, then C is positive definite.

1.9 Let C be an (n, n) positive semidefinite matrix. Show that if s is a vector such that $s'Cs = 0$, then $Cs = 0$. *Hint:* Use Theorem 1.2. Is this result true when C is not positive semidefinite?

1.10 Let C be a (n, n) positive definite matrix and let A be a (m, n) matrix having full row rank. Define $x_0 = (1, 1, \ldots, 1)'$, $u = (1, 1, \ldots, 1)'$ and $b = Ax_0$.

(a) Show how to choose c so that x_0 is optimal for (1.4) with multiplier vector u.

(b) Computationally verify Part (a) by modifying Example1p2.m (Figure 1.4) to solve (1.4) with

$$C = diag(1, 2, 3, 4, 5), \quad A = \begin{bmatrix} 1 & 2 & 3 & 4 & 5 \\ 6 & 7 & 8 & 9 & 10 \end{bmatrix}.$$

1.11 Consider the problem of finding the closest point on the circle $\sum_{i=1}^n (x_i - d_i)^2$ to the line $\sum_{i=1}^n a_i x_i = b$, where $d_1, d_2, \ldots, d_n$ and $a_1, a_2, \ldots, a_n$ are given numbers with $d_i > 0$ for $i = 1, 2, \ldots, n$.

(a) Use Theorem 1.2 to obtain an explicit solution for this problem.

(b) Computationally verify your solution of part (a) by using a modification of Example1p2.m (Figure 1.3) to solve this problem with $d = (1, 2, 3, 4, 5)'$, $a = (1, 1, 2, 2, 4)'$ and $b = 3$.

1.12 Consider the region defined by $x_1 + x_2 \leq 2$ and $2x_1 + 2x_2 \leq 4$. Let $x_0 = (1, 1)'$. How many constraints are active at x_0? Is x_0 an extreme point?

1.13 A degenerate extreme point is an extreme point for which the gradients of those constraints active at it are linearly dependent. Show that every extreme point for the region defined by

$$0 \leq x_i \leq 1, \; i = 1, 2, \ldots, n \; \text{ and } \; \sum_{i=1}^{n} x_i = 1$$

is degenerate.

1.14 Solve the problem of minimizing $-x_1 - x_2$ subject to the constraints of Example 1.2.

Chapter 2

The Efficient Frontier

2.1 The Efficient Frontier

A portfolio consists of various amounts held in different assets. The number of possible assets can be quite large. Standard and Poor's lists 500 assets (S&P 500) and in Canada, the Toronto Stock Exchange lists 300 assets (TSX 300). The basic portfolio optimization problem is to decide how much of an investor's wealth should be optimally invested in each asset.

Consider a universe of n assets. Let μ_i denote the expected return[1] on asset i, $i = 1, 2, \ldots, n$ and σ_{ij} denote the covariance between the returns of assets i and j, $1 \leq i, j \leq n$. Let

$$\mu = (\mu_1, \mu_2, \ldots, \mu_n)' \text{ and } \Sigma = [\sigma_{ij}].$$

Σ is called the covariance matrix[2] for the assets and is symmetric and positive semidefinite. Throughout this chapter we will make the

---

[1]Here, return is defined as the ratio between the price of the i-th asset at any future time and the current price of that i-th asset.

[2]In mathematics, the symbol Σ is used to denote summation. In the finance literature, Σ is used to denote the covariance matrix for risky assets and we shall use that convention throughout this chapter.

stronger assumption that Σ is positive definite. Let x_i denote the proportion of wealth to be invested in asset i and let $x = (x_1, x_2, \ldots, x_n)'$. In terms of x, the expected return of the portfolio μ_p and the variance of the portfolio σ_p^2 are given by

$$\mu_p = \mu'x \text{ and } \sigma_p^2 = x'\Sigma x.$$

Let $l = (1, 1, \ldots, 1)'$; i.e., l is an n-vector of ones. Since the components of x are proportions, they must sum to one; i.e., $l'x = 1$. The constraint $l'x = 1$ is usually called the *budget constraint*.

The goal is to choose a value for x which gives a large value for μ_p and a small value for σ_p^2. These two goals tend to be in conflict. Suppose we have two portfolios, both having the same expected return but the first having a small variance and the second having a large variance. The first portfolio is obviously more attractive because it bears less risk for the same expected return. This is the key idea behind H. Markowitz's definition of an efficient portfolio.

Definition 2.1 *A portfolio is* **variance-efficient** *if for a fixed μ_p, there is no other portfolio which has a smaller variance σ_p^2.*

Definition 2.1 implies that a portfolio is efficient if for some fixed μ_p, σ_p^2 is minimized. Thus the efficient portfolios are solutions of the optimization problem

$$\min\{x'\Sigma x \mid \mu'x = \mu_p, \; l'x = 1\}. \tag{2.1}$$

In (2.1), μ_p is to vary over all possible values. For each value of μ_p, we will in general get a different efficient portfolio. The mathematical structure of (2.1) is that of minimizing a quadratic function subject to two linear equality constraints, the first of which is parametric (the parameter being μ_p). Thus (2.1) is a parametric quadratic programming problem.

There is an alternative (and equivalent) definition of an efficient portfolio. Suppose we have two portfolios both having the same variance but the first having a large expected return and the second having a small expected return. The first portfolio is more attractive because it gives a higher expected return for the same risk as the second portfolio. This prompts the following alternative definition of an efficient portfolio.

Definition 2.2 *A portfolio is* **expected return-efficient** *if for fixed* σ_p^2, *there is no other portfolio with a larger* μ_p.

Definition 2.2 implies that a portfolio is efficient if for some fixed σ_p^2, μ_p is maximized. Using Definition 2.2, the efficient portfolios are the optimal solutions for

$$\max\{\mu'x \mid x'\Sigma x = \sigma_p^2, \ l'x = 1\}. \tag{2.2}$$

Note that (2.2) has a linear objective function, a quadratic equality constraint and a linear equality constraint.

There is a third optimization problem which also produces efficient portfolios. It is a somewhat more convenient formulation than (2.1) or (2.2) and we shall use it to develop the theoretical aspects of portfolio optimization. Let t be a scalar parameter and consider the problem

$$\min\{-t\mu'x + \frac{1}{2}x'\Sigma x \mid l'x = 1\}. \tag{2.3}$$

The intuition behind (2.3) is as follows. For $t \geq 0$, the parameter t balances how much weight is placed on the maximization of $\mu'x$ (equivalently, the minimization of $-\mu'x$) and the minimization of $x'\Sigma x$. If $t = 0$, (2.3) will find the minimum variance portfolio. As t becomes very large, the linear term in (2.3) will dominate and portfolios will be found with higher expected returns at the expense of variance.

Definition 2.3 *A portfolio is* **parametric-efficient** *if it is an optimal solution for (2.3) for some nonnegative parameter t.*

Each of (2.1), (2.2) and (2.3) generates a family of optimal solutions as their respective parameters (μ_p, σ_p^2 and t, respectively) are allowed to vary. A key result is that the families of optimal solutions for the efficient portfolios for each of these problems are identical, provided μ is not a multiple of l. (See Exercise 2.??). Thus we can continue our analysis using any one of the three problems. In what follows, it will be convenient to use (2.3). We next derive a relationship between μ_p and σ_p^2 for efficient portfolios. In doing so, we shall assume that Σ is positive definite and consequently nonsingular (see Exercise 1.7). The approach we use is to obtain the optimal solution for (2.3) as an explicit function of t, use that to obtain expressions for μ_p and σ_p^2 in terms of t, and then eliminate t.

From Definition 1.1 (see Section 1.1), the optimality conditions for (2.3) are:

$$t\mu - \Sigma x = ul, \quad \text{and,} \quad l'x = 1. \tag{2.4}$$

From Theorem 1.2 (see Section 1.1) these conditions are both necessary and sufficient for optimality. Solving the first for x gives

$$x = -u\Sigma^{-1}l + t\Sigma^{-1}\mu. \tag{2.5}$$

Applying the budget constraint gives a single equation for the multiplier u:

$$l'x = 1 = -ul'\Sigma^{-1}l + tl'\Sigma^{-1}\mu,$$

which has solution

$$u = \frac{-1}{l'\Sigma^{-1}l} + t\frac{l'\Sigma^{-1}\mu}{l'\Sigma^{-1}l}.$$

Substituting u into (2.5) gives the efficient portfolios as explicit linear functions of the parameter t:

$$x \equiv x(t) = \frac{\Sigma^{-1}l}{l'\Sigma^{-1}l} + t(\Sigma^{-1}\mu - \frac{l'\Sigma^{-1}\mu}{l'\Sigma^{-1}l}\Sigma^{-1}l). \tag{2.6}$$

Notice that of the quantities in (2.6), $l'\Sigma^{-1}l$ and $l'\Sigma^{-1}\mu$ are both scalars whereas $\Sigma^{-1}l$ and $\Sigma^{-1}\mu$ are both vectors. For ease of notation, let

$$h_0 = \frac{\Sigma^{-1}l}{l'\Sigma^{-1}l} \quad \text{and} \quad h_1 = \Sigma^{-1}\mu - \frac{l'\Sigma^{-1}\mu}{l'\Sigma^{-1}l}\Sigma^{-1}l. \tag{2.7}$$

Then the efficient portfolios are

$$x(t) = h_0 + th_1. \tag{2.8}$$

We next use (2.8) to find μ_p and σ_p^2 in terms of t:

$$\mu_p = \mu'x(t) = \mu'h_0 + t\mu'h_1, \tag{2.9}$$

and

$$\begin{aligned}\sigma_p^2 &= (h_0 + th_1)'\Sigma(h_0 + th_1) \\ &= h_0'\Sigma h_0 + 2th_1'\Sigma h_0 + t^2 h_1'\Sigma h_1\end{aligned} \tag{2.10}$$

and we have used the fact that $h_0'\Sigma h_1 = h_1'\Sigma h_0$ because Σ is symmetric and the transpose of a scalar is just the scalar itself. Let

$$\alpha_0 = \mu'h_0, \quad \alpha_1 = \mu'h_1 \tag{2.11}$$

and

$$\beta_0 = h_0'\Sigma h_0, \quad \beta_1 = h_1'\Sigma h_0 \text{ and } \beta_2 = h_1'\Sigma h_1. \tag{2.12}$$

Note that $\beta_2 > 0$ if and only if μ is not a multiple of l (Exercise 2.1). We will assume throughout that μ is not a multiple of l and this implies

$$\beta_2 > 0. \tag{2.13}$$

The α's and β's are just constants defined in terms of the data for the model problem (2.3). From (2.9), (2.10), (2.11) and (2.12), we now have

$$\mu_p = \alpha_0 + \alpha_1 t \text{ and } \sigma_p^2 = \beta_0 + 2\beta_1 t + \beta_2 t^2, \tag{2.14}$$

which shows μ_p is a linear function of t and σ_p^2 is a quadratic function of t. In addition, it can be shown (Exercise 2.2) that

$$\beta_1 = 0. \tag{2.15}$$

Equation (2.14) gives a parametric relationship between μ_p and σ_p^2. We can eliminate the parameter t and obtain an explicit relationship between μ_p and σ_p^2. Solving (2.14) for t and t^2 and using (2.15) gives

$$t = \frac{(\mu_p - \alpha_0)}{\alpha_1}, \text{ and } t^2 = \frac{(\sigma_p^2 - \beta_0)}{\beta_2}. \tag{2.16}$$

It can be shown (Exercise 2.3) that

$$\beta_2 = \alpha_1. \tag{2.17}$$

Using (2.17) and eliminating t in (2.16) gives

$$\sigma_p^2 - \beta_0 = (\mu_p - \alpha_0)^2/\alpha_1. \tag{2.18}$$

Equation (2.18) shows the relationship between the variance (σ_p^2) and expected return (μ_p) for efficient portfolios and is called the *efficient frontier*. Two of the constants, α_0 and β_0, in (2.18) can be interpreted as follows. When $t = 0$ in (2.3), the problem becomes that of minimizing the variance subject to the budget constraint. The expected return plays no role. The resulting portfolio is called the *global minimum variance portfolio* or simply the *minimum variance portfolio*. From (2.8) we see that h_0 is precisely the minimum variance portfolio and from (2.11), (2.12) and (2.14), α_0 and β_0 are the expected return and variance of the minimum variance portfolio, respectively.

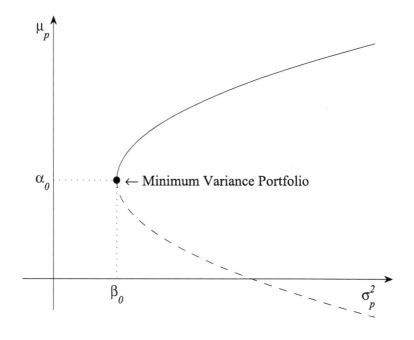

Figure 2.1   The Efficient Frontier.

From (2.18), the algebraic relationship between σ_p^2 (as opposed to σ_p) and μ_p is that of a parabola. The efficient frontier is illustrated in Figure 2.1. The "nose" of the efficient frontier corresponds to the minimum variance portfolio $(t = 0)$ where the investor wants the smallest risk and is not interested in expected return. As t (in (2.16)) is increased from 0, the investor becomes less risk averse and trades off an increase in expected return with increased risk. Thus t is a measure of the investor's aversion to risk.

Points on the efficient frontier below the minimum variance point, correspond to negative values of t. This would correspond to portfolios which are not efficient and therefore only the top half of the efficient frontier is used.

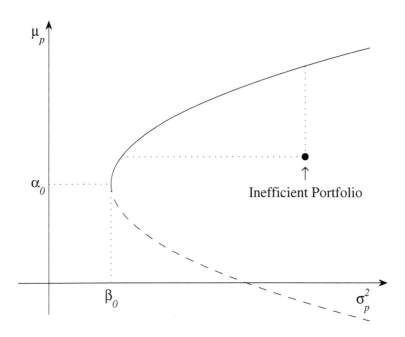

Figure 2.2   Inefficient Portfolio.

Part of the value of the efficient frontier is that it condenses information concerning a problem having n assets into a 2-dimensional graphical representation. Consider the portfolio shown inside the efficient frontier in Figure 2.2. It is inefficient because moving up and in a vertical direction would produce a portfolio with higher expected return at the same level of variance. Similarly, moving to the left in

a horizontal manner would produce a portfolio with decreased risk at the same level of expected return. Thus, all portfolios strictly within the efficient frontier are inefficient. The set of all efficient portfolios corresponds precisely to points on the top half of the efficient frontier.

So far, we have chosen to think of the efficient frontier in (σ_p^2, μ_p) space; i.e., mean-variance space. Sometimes it is helpful to think of it as a curve in (σ_p, μ_p) space; i.e., mean-standard deviation space. Rewriting (2.18) as

$$\sigma_p^2 \; - \; (\mu_p \; - \; \alpha_0)^2/\beta_2 \; = \; \beta_0$$

shows that the efficient frontier depends on the difference of the squares of the two variables μ_p and σ_p. Thus in (σ_p, μ_p) space, the graph of the efficient frontier is a hyperbola. This is illustrated in Figure 2.3.

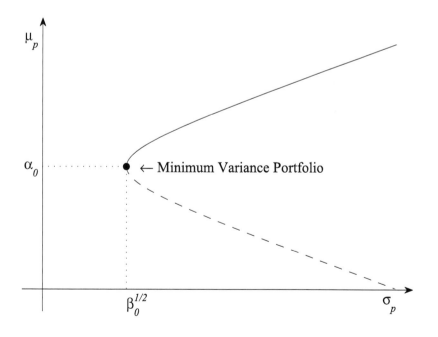

Figure 2.3   Efficient Frontier in Mean-Standard Deviation Space.

We complete this section by illustrating the key concepts with a small numerical example.

Example 2.1

Consider a universe with $n = 3$ assets, $\mu = (1.1, 1.2, 1.3)'$ and $\Sigma = \mathrm{diag}(1 \times 10^{-2}, 5 \times 10^{-2}, 7 \times 10^{-2})$. Find the efficient portfolios, find and sketch the efficient frontier, and computationally verify (2.15) and (2.17) for this problem.

The efficient portfolios are given by (2.8) where h_0 and h_1 are obtained from (2.7). Recall the $l = (1, 1, 1)'$ for this problem. In order to compute h_0, we first must compute $\Sigma^{-1}l$:

$$\Sigma^{-1}l = \begin{bmatrix} 10^2 & 0 & 0 \\ 0 & \frac{1}{5} \times 10^2 & 0 \\ 0 & 0 & \frac{1}{7} \times 10^2 \end{bmatrix} \begin{bmatrix} 1 \\ 1 \\ 1 \end{bmatrix} = 10^2 \begin{bmatrix} 1 \\ \frac{1}{5} \\ \frac{1}{7} \end{bmatrix},$$

from which we compute

$$l'\Sigma^{-1}l = 10^2 \left(1 + \frac{1}{5} + \frac{1}{7}\right) = \frac{47}{35} \times 10^2 = 134.285714.$$

Now we can compute

$$h_0 = \frac{\Sigma^{-1}l}{l'\Sigma^{-1}l} = \frac{35}{47} \times 10^{-2} \times 10^2 \begin{bmatrix} 1 \\ \frac{1}{5} \\ \frac{1}{7} \end{bmatrix} = \frac{35}{47} \begin{bmatrix} 1 \\ \frac{1}{5} \\ \frac{1}{7} \end{bmatrix} = \begin{bmatrix} 0.7446808 \\ 0.1489362 \\ 0.1063830 \end{bmatrix}.$$

Note that $x(0) = h_0 + 0 \cdot h_1 = h_0$ is an efficient portfolio and as such must satisfy the budget constraint; i.e., $l'h_0 = 1$. As a check on our computations, we compute $l'h_0$ for the above h_0:

$$l'h_0 = \frac{35}{47}\left(1 + \frac{1}{5} + \frac{1}{7}\right) = \frac{35}{47} \times \frac{47}{35} = 1,$$

as expected.

We now proceed with the computation of h_1. First we need $\Sigma^{-1}\mu$:

$$\Sigma^{-1}\mu = \begin{bmatrix} 10^2 & 0 & 0 \\ 0 & \frac{1}{5} \times 10^2 & 0 \\ 0 & 0 & \frac{1}{7} \times 10^2 \end{bmatrix} \begin{bmatrix} 1.1 \\ 1.2 \\ 1.3 \end{bmatrix} = 10^2 \begin{bmatrix} 1.1 \\ 0.24 \\ \frac{1.3}{7} \end{bmatrix} = \begin{bmatrix} 110 \\ 24 \\ 18.57143 \end{bmatrix}.$$

We also need $l'\Sigma^{-1}\mu = \mu'\Sigma^{-1}l$:

$$l'\Sigma^{-1}\mu = 10^2(1.1 + 0.24 + \frac{1.3}{7}) = 152.5714286.$$

We can now speed up the computations for h_1 by noting that

$$h_1 = \Sigma^{-1}\mu - l'\Sigma^{-1}\mu \times \left[\frac{\Sigma^{-1}l}{l'\Sigma^{-1}l}\right] = \Sigma^{-1}\mu - (l'\Sigma^{-1}\mu)h_0,$$

and so

$$h_1 = \begin{bmatrix} 110 \\ 24 \\ 18.57143 \end{bmatrix} - 152.5714286 \begin{bmatrix} 0.7446808 \\ 0.1489362 \\ 0.1063830 \end{bmatrix} = \begin{bmatrix} -3.6170135 \\ 1.2765912 \\ 2.3404223 \end{bmatrix}.$$

We have previously noted that $l'h_0 = 1$. Because $x(t) = h_0 + th_1$ is an efficient portfolio for all t, $x(t)$ must also satisfy the budget constraint; i.e., $l'x(t) = l'h_0 + tl'h_1$. But this implies $l'h_1 = 0$. We can use this to check our computations:

$$l'h_1 = (1,1,1) \begin{bmatrix} -3.6170135 \\ 1.2765912 \\ 2.3404223 \end{bmatrix} = 0,$$

(to within hand calculator accuracy) as expected.

Thus, the efficient portfolios for this example are

$$x(t) = \begin{bmatrix} 0.7446808 \\ 0.1489362 \\ 0.1063830 \end{bmatrix} + t \begin{bmatrix} -3.6170135 \\ 1.2765912 \\ 2.3404223 \end{bmatrix}.$$

We next evaluate the efficient set constants $\alpha_0, \alpha_1, \beta_0, \beta_1$ and β_2. From (2.11), we obtain

$$\alpha_0 = \mu'h_0 = (1.1, 1.2, 1.3) \begin{bmatrix} 0.7446808 \\ 0.1489362 \\ 0.1063830 \end{bmatrix} = 1.136170212766,$$

and

$$\alpha_1 = \mu' h_1 = (1.1, 1.2, 1.3) \begin{bmatrix} -3.6170135 \\ 1.2765912 \\ 2.3404223 \end{bmatrix} = 0.595744680851.$$

From (2.12) we obtain

$$
\begin{aligned}
\beta_0 &= h_0' \Sigma h_0 \\
&= (0.7446808, 0.1489362, 0.1063830) \\
&\quad \times \begin{bmatrix} 10^{-2} & 0 & 0 \\ 0 & 5 \times 10^{-2} & 0 \\ 0 & 0 & 7 \times 10^{-2} \end{bmatrix} \begin{bmatrix} 0.7446808 \\ 0.1489362 \\ 0.1063830 \end{bmatrix} \\
&= 0.007446808511,
\end{aligned}
$$

$$
\begin{aligned}
\beta_1 &= h_0' \Sigma h_1 \\
&= (0.7446808, 0.1489362, 0.1063830) \\
&\quad \times \begin{bmatrix} 10^{-2} & 0 & 0 \\ 0 & 5 \times 10^{-2} & 0 \\ 0 & 0 & 7 \times 10^{-2} \end{bmatrix} \begin{bmatrix} -3.6170135 \\ 1.2765912 \\ 2.3404223 \end{bmatrix} \\
&= 0.0,
\end{aligned}
$$

$$
\begin{aligned}
\beta_2 &= h_1' \Sigma h_1 \\
&= (-3.6170135, 1.2765912, 2.3404223) \\
&\quad \times \begin{bmatrix} 10^{-2} & 0 & 0 \\ 0 & 5 \times 10^{-2} & 0 \\ 0 & 0 & 7 \times 10^{-2} \end{bmatrix} \begin{bmatrix} -3.6170135 \\ 1.2765912 \\ 2.3404223 \end{bmatrix} \\
&= 0.595744680851.
\end{aligned}
$$

Note from these results that $\beta_1 = 0$ and $\beta_2 = \alpha_1$ which computationally verifies (2.15) and (2.17), respectively.

From (2.18) the equation of the efficient frontier is

$$\sigma_p^2 - 0.007446808511 = (\mu_p - 1.136170212766)^2 / 0.595744680851.$$

This can be used to construct a table of values for this efficient frontier as shown in Table 2.1. Note that because the efficient frontier is a parabola opening up to the right, each specified value of σ_p^2 will result in two values of μ_p, one on the upper part and one on the lower part. These values are then used to plot the efficient frontier as shown in Figure 2.4.

TABLE 2.1   Table of Values for Efficient Frontier of Example 2.1

σ_p^2	μ_p	μ_p
0.007446808511	1.136170212766	1.136170212766
0.008150764218	1.156648930210	1.115691495322
0.010262631340	1.177127647654	1.095212777878
0.013782409876	1.197606365097	1.074734060435
0.018710099826	1.218085082541	1.054255342991
0.025045701192	1.238563799985	1.033776625547
0.032789213971	1.259042517429	1.013297908103
0.041940638165	1.279521234872	0.992819190659
0.052499973774	1.299999952316	0.972340473216

In all of the numerical examples in this text, we take the components of μ to be "one plus rate of return." The actual rate of return could equally well be used since the efficient portfolios would be identical in either case. Indeed, if μ were replaced with $\mu + \theta l$, where θ is any constant, our model problem

$$\min\{ - t\mu'x + \frac{1}{2}x'\Sigma x \mid l'x = 1\} \tag{2.19}$$

would be changed to

$$\min\{ - t(\mu + \theta l)'x + \frac{1}{2}x'\Sigma x \mid l'x = 1\}. \tag{2.20}$$

However, the budget constraint, $l'x = 1$, implies that the linear part of the objective function for (2.20) can be written

$$
\begin{aligned}
- t(\mu + \theta l)'x &= - t\mu'x - t\theta l'x \\
&= - t\mu'x - t\theta.
\end{aligned}
$$

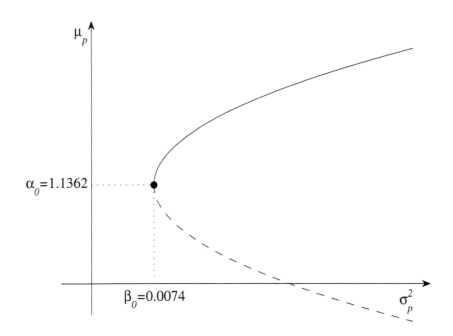

Figure 2.4   Efficient Frontier for Example 2.1.

Since this differs from the objective function for (2.19) by the constant $-t\theta$, the optimal solutions for the two problems will be identical.

2.2   Computer Programs

Figure 2.5 shows the m-file Example2p1.m which solves Example 2.1. Lines 2 and 3 define the data for that problem. Line 4 invokes checkdata (see Figure 1.5) to check if the given covariance matrix is symmetric and positive definite. Line 5 calls the function EFMVcoeff (Efficient Frontier Mean Variance coefficients, see Figure 2.6) with input arguments μ and

Σ. It returns with values for alpha0, alpha1, beta0, beta2, h0 and h1 which represent α_0, α_1, β_0, β_2, h_0 and h_1, respectively. Line 6 calls the function EFMVplot (Efficient Frontier Mean Variance plot, see Figure 2.7) which uses the efficient frontier coefficients just computed to plot the efficient frontier.

```
1  %Example2p1.m
2  mu = [ 1.1 1.2 1.3 ]'
3  Sigma = [1.e-2 0 0 ; 0 5.0e-2 0 ; 0 0 7.0e-2 ]
4  checkdata(Sigma,1.e-6);
5  [alpha0,alpha1,beta0,beta2,h0,h1] = EFMVcoeff(mu,Sigma)
6  EFMVplot(0.0,0.5,0.01,alpha0,alpha1,beta0,beta2);
```

Figure 2.5   Program for Example 2.1: Example2p1.m.

The file EFMVcoeff.m shown in Figure 2.6 calculates the coefficients of the efficient frontier from the input arguments mu and Sigma. Line 4 constructs n from the length of mu and sets ell to be an n-vector of 1's. Line 5 computes temp1 $= \Sigma^{-1}l$ and temp2 $= l'\Sigma^{-1}l$. From this, Line 6 constructs h0 $= (l'\Sigma^{-1}l)^{-1}\Sigma^{-1}l$. Line 7 computes temp3 $= \Sigma^{-1}\mu$ and

```
1   %EFMVcoeff.m
2   function [alpha0,alpha1,beta0,beta2,h0,h1] = ...
3              EFMVcoeff(mu,Sigma)
4   n = length(mu);   ell = ones(1,n)';
5   temp1 = Sigma^-1 * ell;   temp2 = temp1' * ell;
6   h0 = temp2^-1 * temp1
7   temp3= Sigma^-1 * mu;   temp4 = ell' * temp3;
8   h1 = Sigma^-1 * mu - temp4 * h0
9   alpha0 = mu' * h0
10  alpha1 = mu' * h1
11  beta0 = h0' * Sigma * h0
12  beta2 = h1' * Sigma * h1
```

Figure 2.6   Calculation of Efficient Frontier Coefficients: EFMVcoeff.m.

temp4 $= l'\Sigma^{-1}\mu$. Line 8 computes h1 $= \Sigma^{-1}\mu - l'\Sigma^{-1}\mu h_0$. Finally, lines 9 through 12 compute alpha0 $= \mu'h_0$, alpha1 $= \mu'h_1$, beta0 $= h_0'\Sigma h_0$ and beta2 $= h_1'\Sigma h_1$.

Figure 2.7 shows the function EFMVplot. The first three arguments are tlow, thigh and tinc. They specify that the parameter t should start at tlow and increase to thigh in increments of tinc. The last arguments are alpha0, alpha1, beta0 and beta2 which represent the efficient frontier coefficients α_0, α_1, β_0 and β_2, respectively. Lines 5, 6 and 7 compute mup $= \alpha_0 + t\alpha_1$, muplow $= \alpha_0 - t\alpha_1$ (for the lower or inefficient part of the efficient frontier) and the variance sigma2p $= \beta_0 + t^2\beta_2$. Line 8 plots the points, lines 9 and 10 label the axes and line 11 provides a title. Line 12 inserts text showing the minimum variance portfolio. Finally, lines 14 through 18 set up the axes.

```
1  %EFMVplot.m
2  function EFMVplot(tlow,thigh,tinc,alpha0,alpha1, ...
3                         beta0,beta2)
4  t = tlow:tinc:thigh;
5  mup = alpha0 + t * alpha1;
6  muplow = alpha0 - t* alpha1;
7  sigma2p = beta0 + t.^2 * beta2;
8  plot(sigma2p,mup,'-k',sigma2p,muplow,'--k')
9  xlabel('Portfolio Variance \sigma_p^2')
10 ylabel('Portfolio Mean \mu_p')
11 title('Efficient Frontier: Mean-Variance Space',
12 'Fontsize',12)
13 text(beta0,alpha0,'\leftarrow {\rm minimum variance
14 porfolio}')
15 %set axes
16 xmin = 0.;
17 xmax = beta0 + thigh.^2 * beta2;
18 ymin = alpha0 - thigh*alpha1;
19 ymax = alpha0 + thigh*alpha1;
20 axis([xmin xmax ymin ymax])
```

Figure 2.7   Plot the Efficient Frontier: EFMVplot.m.

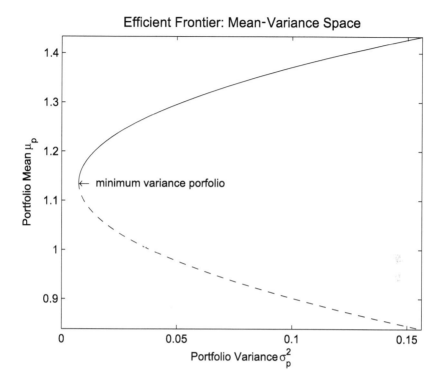

Figure 2.8   Plot of the Efficient Frontier: EfficFrontPlot.m.

2.3   Exercises

2.1 Show $\beta_2 = 0$ if and only if μ is a multiple of l.

2.2 Verify equation (2.15) ($\beta_1 = 0$).

2.3 Verify equation (2.17) ($\beta_2 = \alpha_1$).

2.4 The purpose of this exercise is to establish relationships between the optimal solutions for (2.1), (2.2) and (2.3) and their respective parameters μ_p, σ_p^2 and t.

    (a) Write out the optimality conditions for (2.1) and solve them in terms of Σ^{-1}. Show that these efficient portfolios are identical to those of (2.7)–(2.8). Assume μ is not a multiple of *l. Hints:* (i) (2.1) has *two* linear constraints so that the optimality conditions for it will include *two* multipliers, (ii) it may be helpful to recall a formula for the inverse of a $(2, 2)$ matrix, (iii) μ_p is a parameter in (2.1) and t is a parameter in (2.3). Both parameters can vary over an appropriate range.

(b) Show that the optimal solutions for (2.2) are identical to those for (2.1) and (2.3) for the top part of the efficient frontier; i.e., for $t \geq 0$ in the case of (2.3). What variation of (2.2) will produce the efficient portfolios for the bottom (inefficient) part of the efficient frontier? *Hint:* The optimality conditions for (2.2) can be shown to be

$$(1) \quad x'\Sigma x = \sigma_p^2, \quad l'x = 1,$$

$$(2) \quad \mu = 2\theta_1 \Sigma x + \theta_2 l, \quad \theta_1 \geq 0.$$

2.5 For each of the following problems, find, by hand calculation, $h_0, h_1, x(t), \alpha_0, \alpha_1, \beta_0, \beta_1$ and β_2 (as in Example 2.1), and sketch the efficient frontier.

(a) $n = 2$, $\mu = (1.1, 1.2)'$ and $\Sigma = \text{diag}(10^{-2}, 10^{-1})$,

(b) $n = 3$, $\mu = (1.1, 1.15, 1.2)'$ and $\Sigma = \text{diag}(10^{-4}, 10^{-3}, 10^{-2})$.

In each case, verify your results by using the computer programs of Section 2.2.

2.6 Suppose $\Sigma = \text{diag}(\sigma_i)$. Show that the efficient portfolios have components

$$x_i = [\theta_1 + t(\mu_i - \theta_2)]/\sigma_i, \quad i = 1, 2, \ldots, n,$$

where

$$\theta_1 = 1/(\sigma_1^{-1} + \sigma_2^{-1} + \ldots + \sigma_n^{-1})$$

and

$$\theta_2 = \theta_1(\mu_1/\sigma_1 + \mu_2/\sigma_2 + \ldots + \mu_n/\sigma_n).$$

Also show the multiplier for the budget constraint is $u = -\theta_1 + t\theta_2$.

2.7 Let

$$\mu_{i+1} = (1 + \frac{1}{n-i}) \quad \text{and} \quad \sigma_{i+1}^2 = 10^{-2}(n-i)^2,$$

for $i = 0, 1, \ldots, n-1$ and let $\Sigma = \text{diag}(\sigma_1, \sigma_2, \ldots, \sigma_n)$. Write a MATLAB program to implement the solution for the diagonal covariance case and for each $n = 10$ and $n = 20$ use it to solve the problem with the given data. For each such value of n, also solve the problem using a variation of the program EFMVcoeff.m of Figure 2.6. Compare the results.

2.8 Show the slope of the efficient frontier in (σ_p^2, μ_p) space is $\frac{1}{2t}$ for all $t > 0$.

2.9 Determine the efficient frontier for each of (2.1), (2.2) and (2.3) when μ is a multiple of l; i.e., $\mu = \theta l$ for some scalar θ. Illustrate graphically.

2.10 Consider the problem

$$\min\{ -t\mu'x + \frac{1}{2}x'\Sigma x \mid l'x = 0\},$$

where the data is identical to (2.3) and the problem differs from (2.3) only in that the budget constraint $l'x = 1$ has been replaced by $l'x = 0$. This is usually called a *portfolio rebalance* problem. Obtain a closed form solution for this problem and show that the efficient frontier (in mean-standard deviation space) is a pair of straight lines.

2.11 Suppose $n = 4$,

$$\mu = (1.2, 1.3, 1.4, 1.5)' \quad \text{and} \quad \Sigma = \text{diag}(10^{-4}, 10^{-3}, 10^{-2}, 10^{-1}).$$

(a) Obtain h_0 and h_1.

(b) Find the equation of the efficient frontier. Sketch it.

(c) As t is increased, some asset is eventually reduced to zero. Which one and for what value of t? Assuming that the model problem is now modified by adding nonnegativity constraints $x \geq 0$, and that this asset remains at 0 as t is increased further, find the new optimal portfolio. What is the next asset to be reduced to zero and for what value of t does this occur? Continue until a third asset is reduced to zero. Sketch the resulting four "pieces" of the efficient frontier.

2.12 Is it possible for an efficient portfolio to have one or more components with negative values? These correspond to short selling; i.e., the investor borrows money to buy additional amounts of an attractive asset. Sometimes the portfolio optimization problem is formulated to preclude short selling as we will see in subsequent chapters. The purpose of this exercise is to show that if short selling does occur, the efficient frontier is divided into three segments: the leftmost has portfolios with short selling, the middle has no short selling and the rightmost has portfolios with short selling. See Figure 2.9.

(a) Use (2.8) to show that either there are no nonnegative efficient portfolios or there are numbers t_l and t_h such that $x(t) \geq 0$ for all t with $t_l \leq t \leq t_h$ and $x(t)$ has mixed sign outside of the interval (t_l, t_h).

(b) Use the results of part (a) to show that the nonnegative portfolios all lie on a segment of the efficient frontier, as shown in Figure 2.9.

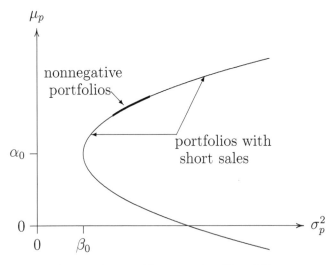

Figure 2.9   Nonnegative Portfolios.

Chapter 3

The Capital Asset Pricing Model

3.1 The Capital Market Line

This section will deal with a variation of the model of the previous section. In addition to the n risky assets, we will now suppose there is an additional asset with special properties. This asset will be risk free and as such will have a zero variance and a zero covariance with the remaining n risky assets. A good example of a risk free asset is Treasury bills or T-bills.

Let x_{n+1} denote the proportion of wealth invested in the risk free asset and let r denote its return. The expected return of this portfolio is

$$
\begin{aligned}
\mu_p &= \mu_1 x_1 + \mu_2 x_2 + \cdots + \mu_n x_n + r x_{n+1} \\
&= (\mu', r) \begin{bmatrix} x \\ x_{n+1} \end{bmatrix}.
\end{aligned}
$$

Its variance is

$$
\sigma_p^2 = x' \Sigma x = \begin{bmatrix} x \\ x_{n+1} \end{bmatrix}' \begin{bmatrix} \Sigma & 0 \\ 0' & 0 \end{bmatrix} \begin{bmatrix} x \\ x_{n+1} \end{bmatrix}.
$$

The covariance matrix for this $n+1$ dimensional problem is

$$\begin{bmatrix} \Sigma & 0 \\ 0' & 0 \end{bmatrix},$$

for which the last row and column contain all zeros corresponding to the risk free asset. This matrix is positive semidefinite, whereas Σ, the covariance matrix for the n risky assets, is assumed to be positive definite. The efficient portfolios are the optimal solutions for (2.3). For the case at hand, (2.3) becomes

$$\left. \begin{array}{ll} \text{minimize}: & -t(\mu', r)\begin{bmatrix} x \\ x_{n+1} \end{bmatrix} + \frac{1}{2}\begin{bmatrix} x \\ x_{n+1} \end{bmatrix}'\begin{bmatrix} \Sigma & 0 \\ 0' & 0 \end{bmatrix}\begin{bmatrix} x \\ x_{n+1} \end{bmatrix} \\ \\ \text{subject to}: & l'x + x_{n+1} = 1 \end{array} \right\}.$$

$$(3.1)$$

In partitioned matrix form, the optimality conditions for (3.1) are

$$t\begin{bmatrix} \mu \\ r \end{bmatrix} - \begin{bmatrix} \Sigma & 0 \\ 0' & 0 \end{bmatrix}\begin{bmatrix} x \\ x_{n+1} \end{bmatrix} = u\begin{bmatrix} l \\ 1 \end{bmatrix}, \qquad (3.2)$$

and

$$l'x + x_{n+1} = 1. \qquad (3.3)$$

The second partition of (3.2) gives the scalar multiplier u:

$$u = tr.$$

Using this, the first partition of (3.2) becomes

$$\Sigma x = t(\mu - rl). \qquad (3.4)$$

Because Σ is positive definite, it is also nonsingular (see Exercise 1.7). Thus, we can obtain the efficient risky assets as

$$x \equiv x(t) = t\Sigma^{-1}(\mu - rl) \qquad (3.5)$$

and the efficient risk free asset

$$x_{n+1} \equiv x_{n+1}(t) = 1 - tl'\Sigma^{-1}(\mu - rl). \qquad (3.6)$$

Using (3.5) and (3.6) we can calculate an efficient portfolio's variance and expected return as follows:

$$\sigma_p^2 = t^2(\mu - rl)'\Sigma^{-1}(\mu - rl), \tag{3.7}$$

$$
\begin{aligned}
\mu_p &= t\mu'\Sigma^{-1}(\mu - rl) + r - trl'\Sigma^{-1}(\mu - rl) \\
&= r + t[\mu'\Sigma^{-1}(\mu - rl) - rl'\Sigma^{-1}(\mu - rl)] \\
&= r + t[(\mu - rl)'\Sigma^{-1}(\mu - rl)]. \tag{3.8}
\end{aligned}
$$

Note that since Σ is positive definite, so too is Σ^{-1} (see Exercise 1.7) and thus

$$(\mu - rl)'\Sigma^{-1}(\mu - rl) > 0$$

provided $\mu \neq rl$. With this restriction, it follows from (3.7) and (3.8) that σ_p^2 and μ_p are both strictly increasing functions of t.

Observe from (3.6) that $x_{n+1}(t)$ is strictly decreasing if and only if $l'\Sigma^{-1}(\mu - rl) > 0$; i.e.,

$$r < \frac{l'\Sigma^{-1}\mu}{l'\Sigma^{-1}l}. \tag{3.9}$$

Because Σ^{-1} is symmetric (Exercise 1.7) and the transpose of a scalar is just the scalar,

$$l'\Sigma^{-1}\mu = (l'\Sigma^{-1}\mu)' = \mu'\Sigma^{-1}l$$

and (3.9) becomes

$$r < \mu'\left[\frac{\Sigma^{-1}l}{l'\Sigma^{-1}l}\right]. \tag{3.10}$$

From (2.7) and (2.11), the quantity on the right-hand side of this inequality is α_0, the expected return of the minimum variance portfolio. Thus we have shown that the allocation in the risk free asset will be a strictly decreasing function of t if and only if the risk free rate is strictly less than the expected return of the minimum variance portfolio. We will assume that $r < \alpha_0$ throughout the remainder of this section.

Because $x_{n+1}(t)$ is a strictly decreasing function of t, as t is increased eventually $x_{n+1}(t)$ is reduced to zero. This occurs when

$$t = t_m \equiv \frac{1}{l'\Sigma^{-1}(\mu - rl)}. \tag{3.11}$$

The corresponding portfolio of risky assets is

$$x(t_m) \equiv x_m = t_m\Sigma^{-1}(\mu - rl). \tag{3.12}$$

x_m is called *the market portfolio* (hence the m subscript). It is that efficient portfolio for which there are zero holdings in the risk free asset. Using (3.5), (3.11) and (3.12) we can write the efficient risky assets as

$$x(t) = \frac{t}{t_m}x_m. \tag{3.13}$$

Using (3.7) and (3.8) we can obtain the equation of the efficient frontier for this model:

$$\mu_p - r = \sigma_p[(\mu - rl)'\Sigma^{-1}(\mu - rl)]^{\frac{1}{2}}. \tag{3.14}$$

In mean-standard deviation space, the efficient frontier is a line. It is called the *Capital Market Line* (CML) and is illustrated in Figure 3.1. Investors move up and down the Capital Market Line according to their aversion to risk. According to (3.5) and (3.6), for $t = 0$ all wealth is invested in the risk free asset and none is invested in the risky assets. As t increases from 0, the amount invested in the risk free asset is reduced whereas the holdings in the risky assets increase. All investors must lie on the CML. From (3.13), the sole thing that distinguishes individual investors is the proportion of the market portfolio they hold.

The CML can be written in a perhaps more revealing way as follows. Recall from (3.5) that part of the optimality conditions for (3.1) are

$$x_m = t_m\Sigma^{-1}(\mu - rl). \tag{3.15}$$

Taking the inner product of both sides with $(\mu - rl)$ gives

$$(\mu - rl)'x_m = t_m(\mu - rl)'\Sigma^{-1}(\mu - rl). \tag{3.16}$$

At the market portfolio, the holdings in the risk free asset have been reduced to zero. Consequently,

$$l'x_m = 1 \quad \text{and} \quad \mu_m = \mu'x_m. \tag{3.17}$$

Using (3.17) in (3.16), taking square roots and rearranging gives

$$\left[\frac{\mu_m - r}{t_m}\right]^{\frac{1}{2}} = [(\mu - rl)'\Sigma^{-1}(\mu - rl)]^{\frac{1}{2}}. \tag{3.18}$$

Now, taking the inner product of both sides of (3.15) with Σx_m gives

$$x'_m \Sigma x_m = t_m(\mu - rl)'x_m = t_m(\mu_m - r),$$

so that

$$\sigma_m = [t_m(\mu_m - r)]^{\frac{1}{2}},$$

or,

$$t_m^{\frac{1}{2}} = \left[\frac{\sigma_m}{(\mu_m - r)^{\frac{1}{2}}}\right]. \tag{3.19}$$

Substitution of (3.19) into (3.18), rearranging and simplifying, gives

$$[(\mu - rl)'\Sigma^{-1}(\mu - rl)]^{\frac{1}{2}} = \frac{(\mu_m - r)}{\sigma_m}. \tag{3.20}$$

Substituting the quantity in the left-hand side of (3.14) shows that the Capital Market Line can also be expressed as

$$\mu_p = r + \left[\frac{(\mu_m - r)}{\sigma_m}\right]\sigma_p. \tag{3.21}$$

Equation (3.21) shows that the slope of the CML is

$$\frac{(\mu_m - r)}{\sigma_m}.$$

This is in agreement with direct observation in Figure 3.1.

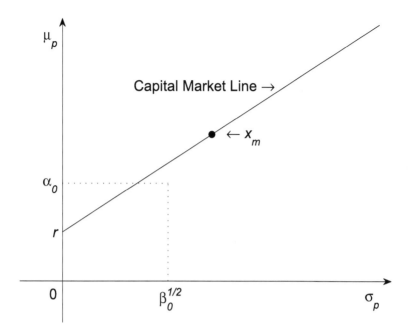

Figure 3.1   Capital Market Line.

Consider the following implication of this model. Suppose in an economy that everyone is mean-variance efficient and everyone "sees" the same covariance matrix Σ and the same vector of expected returns μ. Let two typical investors be Mr. X and Ms. Y and suppose their t values are t_x and t_y, respectively. Note that t_x and t_y could be obtained from the amounts Mr. X and Ms. Y invest in the risk free asset by means of (3.6). Let Mr. X and Ms. Y's holdings in the risky assets be denoted by x and y, respectively. From (3.13),

$$x = \frac{t_x}{t_m} x_m, \quad \text{and,} \quad y = \frac{t_y}{t_m} x_m.$$

This implies that their holdings in risky assets are *proportional*; i.e., $x = (t_x/t_y)y$. For example, suppose $n = 4$ and Mr. X's risky holdings are $x = (0.1, 0.11, 0.12, 0.13)'$. Suppose we knew Ms. Y's holdings in just the first risky asset were 0.2. But then her holdings in the remaining risky assets must be in the same proportion (namely 2) as Mr. X's. Thus we would know the remainder of her holdings: $y = (0.2, 0.22, 0.24, 0.26)'$.

We next turn to the question of what happens when t exceeds t_m. By definition, if the Capital Market Line were to continue, holdings in the risk free asset would become negative. This would allow short selling the risk free asset, which we will not allow in this model. For all $t \geq t_m$, we shall impose the condition $x_{n+1}(t) = 0$. Substituting $x_{n+1} = 0$ everywhere in (3.1), the problem (3.1) reduces to (2.3), the n risky asset problem of the previous section for $t \geq t_m$. Figure 3.2 shows the Capital Market Line and the efficient frontier for the n risky asset problem superimposed, in mean-standard deviation space.

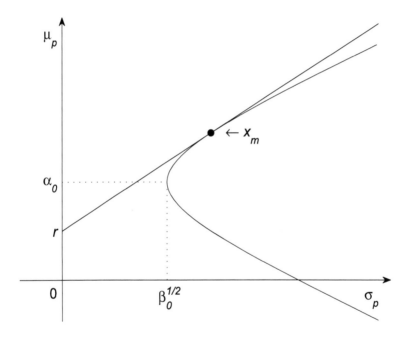

Figure 3.2   Capital Market Line and Efficient Frontier for Risky Assets.

Implicit in the model problem (3.1) is the assumption that the risk free asset must be nonnegative. The key result for this problem is that the efficient frontier for it is composed of two pieces (see Figure 3.3). The first piece is the Capital Market Line going from r to the market portfolio x_m. The second piece is a part of the hyperbola (2.18), the efficient frontier for the risky assets. The two pieces of the efficient frontier are shown as the solid line in Figure 3.3.

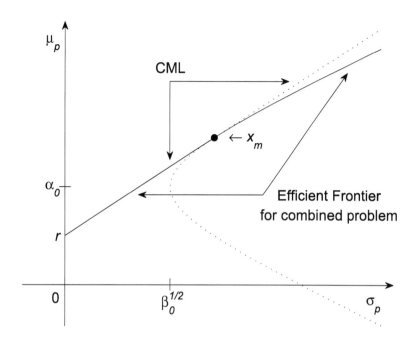

Figure 3.3   Two Pieces of the Efficient Frontier.

Figure 3.3 is drawn in a special way, and not by accident. The mean
and standard deviation at x_m on the CML is identical to that for the
risky asset part of the frontier. This is always the case. Indeed, the two
corresponding portfolios are both x_m. To see this, from (2.8) and (3.11)
we can calculate

$$
\begin{aligned}
x(t_m) &= \frac{\Sigma^{-1}l}{l'\Sigma^{-1}l} + \frac{1}{l'\Sigma^{-1}(\mu - rl)}\left[\Sigma^{-1}\mu - \frac{l'\Sigma^{-1}\mu}{l'\Sigma^{-1}l}\Sigma^{-1}l\right] \\
&= \left[1 - \frac{l'\Sigma^{-1}\mu}{l'\Sigma^{-1}(\mu - rl)}\right]\frac{\Sigma^{-1}l}{l'\Sigma^{-1}l} + \frac{1}{l'\Sigma^{-1}(\mu - rl)}\Sigma^{-1}\mu \\
&= \frac{-rl'\Sigma^{-1}l}{l'\Sigma^{-1}(\mu - rl)}\frac{\Sigma^{-1}l}{l'\Sigma^{-1}l} + \frac{\Sigma^{-1}\mu}{l'\Sigma^{-1}(\mu - rl)} \\
&= \frac{\Sigma^{-1}(\mu - rl)}{l'\Sigma^{-1}(\mu - rl)} \\
&= x_m.
\end{aligned}
$$

Since the two portfolios are identical, their means are equal and so too
are their standard deviations. Thus, the efficient frontier is continuous.

Not only is the efficient frontier continuous, it is also differentiable. One way to think of this is that in mean-standard deviation space, the CML is tangent to the efficient frontier for the risky assets. This is shown in Figure 3.3. From (3.14), the slope of the CML is

$$((\mu - rl)'\Sigma^{-1}(\mu - rl))^{\frac{1}{2}}. \tag{3.22}$$

To verify the differentiability assertion, we need only show that the slope of the efficient frontier for risky assets at x_m has this same value. A convenient way to do this is to use (2.14), namely

$$\mu_p = \alpha_0 + \alpha_1 t \text{ and } \sigma_p = (\beta_0 + 2\beta_1 t + \beta_2 t^2)^{\frac{1}{2}}.$$

Using the chain rule of calculus, $\beta_1 = 0$ (2.15) and $\beta_2 = \alpha_1$ (2.17), we determine

$$\frac{d\mu_p}{d\sigma_p} = \frac{d\mu_p}{dt}\frac{dt}{d\sigma_p}$$

$$= \alpha_1 \left[\frac{2\beta_2 t}{2\sigma_p}\right]^{-1}$$

$$= \frac{\sigma_p}{t}. \tag{3.23}$$

When $t = t_m$, (3.7) with (3.12) imply

$$\sigma_p = \left[\frac{(\mu - rl)'\Sigma^{-1}(\mu - rl)}{(l'\Sigma^{-1}(\mu - rl))^2}\right]^{\frac{1}{2}}, \tag{3.24}$$

so that with (3.11) we have

$$\frac{d\mu_p}{d\sigma_p} = ((\mu - rl)'\Sigma^{-1}(\mu - rl))^{\frac{1}{2}}, \tag{3.25}$$

and this is identical to (3.22). Thus the slope of the CML is equal to the slope of the efficient frontier for risky assets at x_m and the total efficient frontier is indeed differentiable.

Example 3.1

Find the equation of the CML using (3.14), t_m and the market portfolio x_m for the problem of Example 2.1 and a risk free return of $r = 1.02$. Verify the equation of the CML by recalculating it using (3.21).

The data for Example 2.1 has $n = 3$, $\mu = (1.1, 1.2, 1.3)'$ and $\Sigma = \text{diag}(10^{-2}, 5 \times 10^{-2}, 7 \times 10^{-2})$. The required quantities all depend on $\Sigma^{-1}(\mu - rl)$ so we calculate this first:

$$\Sigma^{-1}(\mu - rl) = 100 \begin{bmatrix} 1 & 0 & 0 \\ 0 & \frac{1}{5} & 0 \\ 0 & 0 & \frac{1}{7} \end{bmatrix} \begin{bmatrix} 0.08 \\ 0.18 \\ 0.28 \end{bmatrix} = \begin{bmatrix} 8 \\ 3.6 \\ 4 \end{bmatrix} . \quad (3.26)$$

Thus

$$(\mu - rl)'\Sigma^{-1}(\mu - rl) = \begin{bmatrix} 0.08 & 0.18 & 0.28 \end{bmatrix} \begin{bmatrix} 8 \\ 3.6 \\ 4 \end{bmatrix} = 2.4080$$

and since $\sqrt{2.4080} = 1.5518$, it follows from (3.14) that the equation of the CML is

$$\mu_p = 1.02 + 1.5518\, \sigma_p. \quad (3.27)$$

From (3.11), we have

$$t_m = \frac{1}{l'\Sigma^{-1}(\mu - rl)}.$$

From (3.26),

$$l'\Sigma^{-1}(\mu - rl) = \begin{bmatrix} 1 & 1 & 1 \end{bmatrix} \begin{bmatrix} 8 \\ 3.6 \\ 4 \end{bmatrix} = 15.6,$$

so that

$$t_m = 0.0641.$$

Finally, we have from (3.12) and (3.26) that the market portfolio is

$$\begin{aligned} x_m &= t_m \Sigma^{-1}(\mu - rl) \\ &= 0.0641 \begin{bmatrix} 8 \\ 3.6 \\ 4 \end{bmatrix} \\ &= \begin{bmatrix} 0.5128 \\ 0.2308 \\ 0.2564 \end{bmatrix} . \end{aligned}$$

We next calculate the equation of the CML using (3.21) as follows. Using x_m above, we calculate

$$\mu_m = \mu' x_m = 1.1744, \qquad \sigma_m^2 = x_m' \Sigma x_m = 0.0099,$$

and

$$\sigma_m = 0.0995, \qquad \frac{\mu_m - r}{\sigma_m} = \frac{1.1744 - 1.02}{0.0995} = 1.5518.$$

From (3.21), the equation of the CML is

$$\mu_p = 1.02 + 1.5518\,\sigma_p,$$

in agreement with (3.27). ◇

3.2 The Security Market Line

In this section we continue using the model problem Section 3.1 and repeat that model problem here for convenience:

$$\left. \begin{array}{l} \text{minimize}: \quad -t(\mu', r)\begin{bmatrix} x \\ x_{n+1} \end{bmatrix} + \frac{1}{2}\begin{bmatrix} x \\ x_{n+1} \end{bmatrix}'\begin{bmatrix} \Sigma & 0 \\ 0' & 0 \end{bmatrix}\begin{bmatrix} x \\ x_{n+1} \end{bmatrix} \\[2em] \text{subject to}: \qquad\qquad l'x + x_{n+1} = 1 \end{array} \right\}.$$

$$(3.28)$$

The optimality conditions for (3.28) were formulated in Section 3.1 as (3.4) and (3.3). We repeat these here.

$$\Sigma x = t(\mu - rl), \qquad\qquad (3.29)$$

$$l'x + x_{n+1} = 1. \qquad\qquad (3.30)$$

We now let $t = t_m$. It follows that $x = x_m$, $x_{n+1} = 0$ and $l'x_m = 1$. Equation (3.29) now becomes

$$t_m\mu = \Sigma x_m + t_m rl,$$

or,

$$\mu = rl + \frac{1}{t_m}\Sigma x_m. \tag{3.31}$$

Taking the inner product of both sides of (3.31) with x_m gives

$$\mu'x_m = rl'x_m + \frac{1}{t_m}x'_m\Sigma x_m. \tag{3.32}$$

But $\mu'x_m = \mu_m$, $l'x_m = 1$ and $x'_m\Sigma x_m = \sigma_m^2$, so (3.32) becomes

$$\mu_m = r + \frac{\sigma_m^2}{t_m}.$$

Upon rearrangement, this becomes

$$\frac{\mu_m - r}{\sigma_m^2} = \frac{1}{t_m}. \tag{3.33}$$

Substituting (3.33) into (3.31) gives

$$\mu = rl + \left[\frac{\mu_m - r}{\sigma_m^2}\right]\Sigma x_m. \tag{3.34}$$

Now let r_i denote the random variable whose value is the return from asset i, and $E(r_i) = \mu_i$, $i = 1, 2, \ldots, n$. That is, μ_i is the expected value of r_i. The i-th component of (3.34) can be written

$$\mu_i - r = \left[\frac{\mu_m - r}{\sigma_m^2}\right][\sigma_{i1}(x_m)_1 + \sigma_{i2}(x_m)_2 + \cdots + \sigma_{in}(x_m)_n]. \tag{3.35}$$

But

$$[\sigma_{i1}(x_m)_1 + \sigma_{i2}(x_m)_2 + \cdots + \sigma_{in}(x_m)_n]$$

is just the covariance of r_i with the market portfolio x_m which has return

$$r_m = (x_m)_1 r_1 + (x_m)_2 r_2 + \cdots + (x_m)_n r_n$$

and satisfies

$$(x_m)_1 + (x_m)_2 + \cdots + (x_m)_n = 1.$$

Thus, (3.35) can be rewritten as

$$\mu_i = r + \left[\frac{\mu_m - r}{\sigma_m^2}\right] \text{cov}(r_i, r_m) \quad i = 1, 2, \ldots, n. \tag{3.36}$$

Equation (3.36) is called the *Security Market Line (SML)* and shows that the expected returns of all assets must plot on a line with slope

$$\left[\frac{\mu_m - r}{\sigma_m^2}\right]$$

and μ intercept r, in $(\text{cov}(r_i, r_m), E(r_i))$ space.

The market portfolio also lies on the SML. See Exercise 3.6.

Example 3.2
For the problem of Example 3.1, find the equation of the SML and numerically verify that μ_1, μ_2 and μ_3 lie on it.

From the results of Example 3.1, we have $\mu_m = 1.1744$ and $\sigma_m = 0.0995$ so that the slope of the SML is

$$\frac{\mu_m - r}{\sigma_m^2} = \frac{1.1744 - 1.02}{(0.0995)^2} = 15.5999.$$

We can write the Security Market Line in vector format:

$$\mu = rl + \left[\frac{\mu_m - r}{\sigma_m^2}\right] \Sigma x_m,$$

$$= \begin{bmatrix} 1.02 \\ 1.02 \\ 1.02 \end{bmatrix} + 15.5999 \Sigma x_m. \tag{3.37}$$

We can also write the SML in component form as

$$\mu_i = r + 15.5999 \times \text{cov}(r_i, r_m),$$

for $i = 1, 2, \ldots, n$. It is straightforward to compute

$$\Sigma x_m = \begin{bmatrix} 0.0051 \\ 0.0115 \\ 0.0179 \end{bmatrix}.$$

Inserting this last into the right-hand side of (3.37) and simplifying gives in agreement with the left-hand side (namely μ) and so the μ_i, $i = 1, 2, 3$ do indeed plot on the SML. See Figure 3.4.   ◇

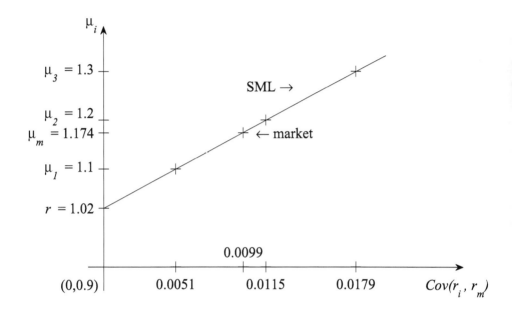

Figure 3.4   SML for Example 3.2.

3.3   Computer Programs

Figure 3.5 shows the m-file Example3p1.m which is the main routine for plotting the CML for Example 3.1. Lines 2–4 set up the data. Line 5 defines the lower, upper and increment on t for plotting the CML using (3.7) and (3.8). Line 6 performs the validity check on the covariance matrix. Line 7 calls the function EFMVcoeff (Figure 2.6) to obtain the coefficients of the efficient frontier for the risky assets. Lines 8 and 9 call CMLplot (Figure 3.6) to plot the CML.

```
1  %Example3p1.m
2  mu = [ 1.1 1.2 1.3 ]'
3  Sigma = [1.e-2 0 0 ; 0 5.0e-2 0 ; 0 0 7.0e-2 ]
4  r = 1.02
5  tlow=0.; thigh=1.; tinc=0.01;
6  checkdata(Sigma,1.e-6);
7  [alpha0,alpha1,beta0,beta2,h0,h1] = EFMVcoeff(mu,Sigma)
8  CMLplot(tlow,thigh,tinc,alpha0,alpha1,beta0,...
9                    beta2,mu,Sigma,r);
```

Figure 3.5   Example3p1.m.

Figure 3.6 shows the m-file CMLplot.m which plots the Capital Market Line. Its input parameters are the limits on t for plotting the CML, the coefficients of the efficient frontier and the problem data μ, Σ and r. Line 4 calls EFMSDplot (Efficient Frontier Mean-Standard Deviation plot) to plot the efficient frontier in mean-standard deviation space (see Figure 3.8). Line 5 keeps the current plot open for further additions. Line 8 calculates the slope of the CML, namely $[(\mu - rl)'\Sigma^{-1}(\mu - rl)]^{1/2}$. Lines 9 and 10 compute t_m and x_m, respectively. Lines 11–13 compute points on the CML and line 14 plots them. Lines 15 through 18 plot the extension of the CML using a dashed line. Lines 19 through 25 label some key points in the graph.

```
1   %CMLplot.m
2   function CMLplot(tlow,thigh,tinc,alpha0,alpha1,...
3                           beta0,beta2,mu,Sigma,r)
4   EFMSDplot(tlow,thigh,tinc,alpha0,alpha1,beta0,beta2);
5   hold on
6   n = length(mu);   ell = ones(1,n)';
7   coeff = (mu − r*ell)' * Sigma^−1 * (mu − r*ell)
8   coeffsqrt = coeff.^0.5
9   tm = 1./(ell' * Sigma^−1 * (mu − r*ell))
10  xm = tm *Sigma^−1 * (mu − r*ell)
11  t = tlow:tinc.^2:tm;
12  mup = r + t * coeff;
13  sigmap = t* coeffsqrt;
14  plot(sigmap,mup,'k')
15  t = tm:tinc.^2:3*tm;
16  mup = r + t * coeff;
17  sigmap = t* coeffsqrt;
18  plot(sigmap,mup,'k−.')
19  mup = r + tm * coeff;   sigmap = tm*coeffsqrt;
20  xlabel('Portfolio Standard Deviation \sigma_p')
21  ylabel('Portfolio Mean \mu_p')
22  text(sigmap,mup,...
23  '\leftarrow {\rm market porfolio}, t = t_m')
24  text(sigmap/2.,(mup+r)/2.,...
25  '\leftarrow {\rm Capital Market Line}')
26  text(2*sigmap,r+2*tm*coeff,...
27  '\leftarrow {\rm Extension of Capital Market Line}')
```

Figure 3.6   CMLplot.m.

Figure 3.7 shows EFMSDplot.m which plots the efficient frontier for the risky assets in mean-standard deviation space. Lines 4 through 6 compute points on the efficient frontier and its lower extension and plots them. Lines 9 through 13 construct the axes. Figure 3.8 shows the final plot.

```
1  %EFMSDplot.m
2  function EFMSDplot(tlow,thigh,tinc,alpha0,...
3                           alpha1,beta0,beta2)
4  t = tlow:tinc:thigh;
5  mup = alpha0 + t * alpha1;
6  muplow = alpha0 - t* alpha1;
7  stddevp = (beta0 + t.^2 * beta2).^0.5;
8  plot(stddevp,mup,'k',stddevp,muplow,'k--')
9  xmin = 0.;
10 xmax = (beta0 + thigh.^2 * beta2).^0.5;
11 ymin = alpha0 - thigh*alpha1;
12 ymax = alpha0 + thigh*alpha1;
13 axis([xmin xmax ymin ymax])
```

Figure 3.7   EFMSDplot.m.

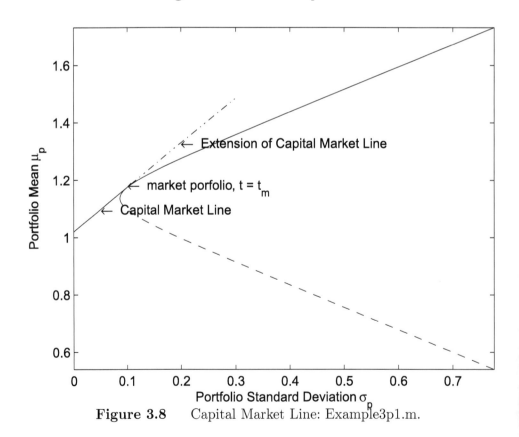

Figure 3.8   Capital Market Line: Example3p1.m.

3.4   Exercises

3.1 Each of the following problems is to be augmented with a risk free asset with return $r = 1.05$. Determine the efficient portfolios and the two pieces of the efficient frontier.

(a) $n = 2$, $\mu = (1.1, 1.2)'$ and $\Sigma = \text{diag}(10^{-2}, 10^{-1})$,

(b) $n = 3$, $\mu = (1.1, 1.15, 1.2)'$ and $\Sigma = \text{diag}(10^{-4}, 10^{-3}, 10^{-2})$.

3.2 Suppose the data for problem (2.11) (with no nonnegativity constraints) is augmented by a risk free asset with $r = 1.1$. Determine all the efficient portfolios, x_m and draw the two pieces of the efficient frontier in (σ_p, μ_p) space.

3.3 Suppose $\Sigma = \text{diag}(\sigma_1, \sigma_2, \ldots, \sigma_n)$ and $\sigma_i > 0$, $i = 1, 2, \ldots, n$. Show that on the CML each investor i has holdings proportional to

$$\frac{\mu_i - r}{\sigma_i}, \quad i = 1, 2, \ldots, n.$$

Obtain an explicit formula for x_m.

3.4 Consider a universe of two assets. The first asset is risky and has an expected return and variance of 1.2 and 10^{-2}, respectively. The second asset is risk free and has a return of 1.1.

(a) Find the equations of the two pieces of the efficient frontier.

(b) Explain your results in part (a) in terms of any assumptions that may or may not have been made.

3.5 Find the SML for each of the problems of Exercise 3.1. Also plot the SML as in Figure 3.4.

3.6 Show that the market portfolio lies on the SML.

Chapter 4

Sharpe Ratios and Implied Risk Free Returns

The typical situation studied in Chapter 3 is illustrated in Figure 4.1. The key quantities are the implied risk free return r_m, the tangency point of the CML with the efficient frontier defined by μ_m and the slope, R, of the CML. It is apparent from Figure 4.1 that if any single one of r_m, μ_m or R is changed then the other two will change accordingly. It is the purpose of this chapter to determine precisely what these changes will be.

In this chapter, we will look at the following problems.

1. Given the expected return on the market portfolio μ_m, what is the implied risk free return r_m?

2. Given the implied risk free return r_m, what is the implied market portfolio x_m or equivalently, what is μ_m or t_m from which all other relevant quantities can be deduced?

3. Given the slope R of the CML, what is the implied market portfolio x_m or equivalently, what is μ_m or t_m from which all other relevant quantities can be deduced?

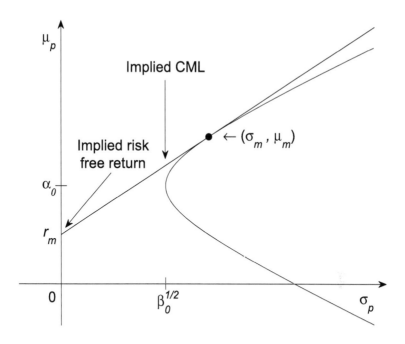

Figure 4.1   Risk Free Return and CML Implied by a Market Portfolio.

This analysis is important because if an analyst hypothesizes that one of μ_m, r_m or R is changed, the effect on the remaining two parameters will immediately be known.

In Section 4.1, we will perform this analysis for the basic Markowitz model with just a budget constraint. In Section 4.2, we will provide a similar analysis from a different point of view: one which will allow generalization of these results to problems which have general linear inequality constraints.

4.1   Direct Derivation

In this section we will look at problems closely related to the CML of the previous chapter but from an opposing point of view. We will continue analyzing the model problem

$$\text{minimize}: \quad -t(\mu', r)\begin{bmatrix} x \\ x_{n+1} \end{bmatrix} + \tfrac{1}{2}\begin{bmatrix} x \\ x_{n+1} \end{bmatrix}'\begin{bmatrix} \Sigma & 0 \\ 0' & 0 \end{bmatrix}\begin{bmatrix} x \\ x_{n+1} \end{bmatrix} \Bigg\}.$$

$$\text{subject to}: \qquad l'x + x_{n+1} = 1, \quad x_{n+1} \geq 0$$

$$\tag{4.1}$$

Note that we have made explicit the nonnegativity constraint on the risk free asset.

Consider the efficient frontier for the risky assets. Let α_0, α_1, β_0 and β_2 be the parameters which define this efficient frontier. Suppose we pick a point on it with coordinates (σ_m, μ_m) satisfying $\mu_m > \alpha_0$. Next we require that point to correspond to the market portfolio. The corresponding CML must be tangent to the efficient frontier at this point. We can determine this line as follows. The efficient frontier for the risky assets (2.18) can be rewritten as

$$\mu_p = \alpha_0 + (\alpha_1(\sigma_p^2 - \beta_0))^{\frac{1}{2}}, \tag{4.2}$$

from which

$$\frac{d\mu_p}{d\sigma_p} = \frac{\alpha_1 \sigma_p}{(\alpha_1(\sigma_p^2 - \beta_0))^{\frac{1}{2}}}. \tag{4.3}$$

At (σ_m, μ_m),

$$\frac{d\mu_p}{d\sigma_p} = \frac{\alpha_1 \sigma_m}{(\alpha_1(\sigma_m^2 - \beta_0))^{\frac{1}{2}}}.$$

The tangent line to the efficient frontier at (σ_m, μ_m) is thus

$$\mu_p = \mu_m + \frac{\alpha_1 \sigma_m}{(\alpha_1(\sigma_m^2 - \beta_0))^{\frac{1}{2}}}(\sigma_p - \sigma_m).$$

We have thus shown the following result:

Lemma 4.1 *Let (σ_m, μ_m) be any point on the efficient frontier with $\mu_m > \alpha_0$. Then the equation of the line tangent to the efficient frontier at that point is*

$$\mu_p = \mu_m + \frac{\alpha_1 \sigma_m}{(\alpha_1(\sigma_m^2 - \beta_0))^{\frac{1}{2}}}(\sigma_p - \sigma_m).$$

We next use Lemma 4.1 to prove the key results of this section.

Theorem 4.1 *(a) Let the expected return on the market portfolio be μ_m with $\mu_m > \alpha_0$. Then the implied risk free return is*

$$r_m = \alpha_0 - \frac{\alpha_1 \beta_0}{\mu_m - \alpha_0} \text{ with } t_m = \frac{\mu_m - \alpha_0}{\alpha_1}.$$

(b) Let the risk free return be r_m with $r_m < \alpha_0$. Then the expected return of the implied market portfolio is given by

$$\mu_m = \alpha_0 + \frac{\alpha_1 \beta_0}{\alpha_0 - r_m} \text{ with } t_m = \frac{\beta_0}{\alpha_0 - r_m}.$$

(c) Let R be the given slope of the CML with $R > \sqrt{\beta_2}$. Then the implied risk free return r_m is given by

$$r_m = \alpha_0 - [\beta_0(R^2 - \beta_2)]^{1/2} \text{ with } t_m = \frac{\beta_0^{1/2}}{(R^2 - \beta_2)^{1/2}}.$$

Proof:
(a) r_m is the intercept of the line specified in Lemma 4.1, with $\sigma_p = 0$. Thus

$$r_m = \mu_m - \frac{\alpha_1 \sigma_m^2}{(\alpha_1(\sigma_m^2 - \beta_0))^{\frac{1}{2}}}. \tag{4.4}$$

But (σ_m, μ_m) lies on the efficient frontier (4.2); i.e.,

$$\mu_m = \alpha_0 + (\alpha_1(\sigma_m^2 - \beta_0))^{\frac{1}{2}}.$$

Substitution of this into (4.4) gives the intercept

$$\begin{aligned}
r_m &= \alpha_0 + (\alpha_1(\sigma_m^2 - \beta_0))^{\frac{1}{2}} - \frac{\alpha_1 \sigma_m^2}{(\alpha_1(\sigma_m^2 - \beta_0))^{\frac{1}{2}}} \\
&= \alpha_0 + \frac{\alpha_1(\sigma_m^2 - \beta_0) - \alpha_1 \sigma_m^2}{(\alpha_1(\sigma_m^2 - \beta_0))^{\frac{1}{2}}} \\
&= \alpha_0 - \frac{\alpha_1 \beta_0}{\mu_m - \alpha_0}.
\end{aligned}$$

Noting that $\mu_m = \alpha_0 + \alpha_1 t_m$ and solving for t_m gives

$$t_m = \frac{\mu_m - \alpha_0}{\alpha_1},$$

which completes the proof of part (a).

(b) Solving for μ_m in part (a) gives

$$\mu_m = \alpha_0 + \frac{\alpha_1 \beta_0}{\alpha_0 - r_m},$$

$$= \alpha_0 + \alpha_1 \left[\frac{\beta_0}{\alpha_0 - r_m} \right], \qquad (4.5)$$

which verifies the desired result for μ_m. Recall that $\mu_m = \alpha_0 + \alpha_1 t_m$. Comparing this formula with (4.5) it follows that

$$t_m = \frac{\beta_0}{\alpha_0 - r_m},$$

as required.

(c) From Lemma 4.1, the slope of the CML is

$$\frac{\alpha_1 \sigma_m}{(\alpha_1 (\sigma_m^2 - \beta_0))^{\frac{1}{2}}}.$$

Equating this to R, squaring and solving for σ_m^2 gives

$$\sigma_m^2 = \frac{R^2 \beta_0}{R^2 - \alpha_1}.$$

But $\sigma_m^2 = \beta_0 + t_m^2 \beta_2$. Equating the two expressions for σ_m^2 and using the fact that $\beta_2 = \alpha_1$ gives

$$t_m = \frac{\beta_0^{1/2}}{(R^2 - \beta_2)^{1/2}}.$$

Now $\mu_m = \alpha_0 + t_m \alpha_1$ so

$$\mu_m = \alpha_0 + \frac{\alpha_1 \beta_0^{1/2}}{(R^2 - \beta_2)^{1/2}}.$$

We now use this expression for μ_m in the expression for the implied risk free return in part (a). This gives

$$
\begin{aligned}
r_m &= \alpha_0 - \frac{\alpha_1 \beta_0}{\mu_m - \alpha_0} \\
&= \alpha_0 - \frac{\alpha_1 \beta_0 (R^2 - \beta_2)^{1/2}}{\alpha_1 \beta_0^{1/2}} \\
&= \alpha_0 - [\beta_0 (R^2 - \beta_2)]^{1/2},
\end{aligned}
$$

and this completes the proof of the theorem.   □

In Theorem 4.1(a) recall from (2.12), (2.13) and (2.17) that $\alpha_1 > 0$ and $\beta_0 > 0$. It then follows that $r_m < \alpha_0$ and that r_m tends to α_0 as μ_m tends to infinity. See Figure 4.1.

We illustrate Theorem 4.1 in the following example.

Example 4.1

For the given data, answer the following questions.

(a) Find the risk free return r_m, implied by a market expected return of $\mu_m = 1.4$. Also find the associated t_m, μ_m and σ_m^2. What market expected return is implied by r_m?

(b) Suppose the risk free return is changed from $r_m = 1.050$ to $r_m = 1.055$. What is the change in the expected return on the market portfolio?

(c) If the risk free return is $r_m = 1.05$, what is the market portfolio?

The data are

$$
\mu = \begin{bmatrix} 1.1 \\ 1.15 \\ 1.2 \\ 1.27 \\ 1.3 \end{bmatrix}, \quad
\Sigma = \begin{bmatrix}
0.0100 & 0.0007 & 0 & 0 & 0.0006 \\
0.0007 & 0.0500 & 0.0001 & 0 & 0 \\
0 & 0.0001 & 0.0700 & 0 & 0 \\
0 & 0 & 0 & 0.0800 & 0 \\
0.0006 & 0 & 0 & 0 & 0.0900
\end{bmatrix}.
$$

The efficient set coefficients could be found by hand calculation. However, this would require inverting the $(5, 5)$ covariance matrix Σ which is quite tedious by hand. Rather, we use the computer program "EFMV-coeff.m" (Figure 2.6) to do the calculations. Running the program with the given data produces

$$h_0 = \begin{bmatrix} 0.6373 \\ 0.1209 \\ 0.0927 \\ 0.0812 \\ 0.0680 \end{bmatrix}, \quad h_1 = \begin{bmatrix} -4.3919 \\ 0.2057 \\ 0.8181 \\ 1.5911 \\ 1.7770 \end{bmatrix},$$

and

$$\alpha_0 = 1.1427, \quad \alpha_1 = 0.7180, \quad \beta_0 = 0.0065, \quad \beta_2 = 0.7180.$$

(a) We can use Theorem 4.1(a) to calculate

$$r_m = \alpha_0 - \frac{\alpha_1 \beta_0}{\mu_m - \alpha_0} = 1.1427 - \frac{0.7180 \times 0.0065}{1.4 - 1.1427} = 1.1246.$$

We also calculate

$$t_m = \frac{\mu_m - \alpha_0}{\alpha_1} = \frac{1.4 - 1.1427}{0.7180} = 0.3584.$$

Furthermore, with this value of r_m we can use Theorem 4.1(b) to calculate

$$\mu_m = \alpha_0 + \frac{\beta_0 \alpha_1}{\alpha_0 - r_m} = 1.4,$$

as expected. Finally, we calculate σ_p^2 according to

$$\sigma_p^2 = \beta_0 + t_m^2 \beta_2 = 0.0987.$$

(b) We use Theorem 4.1(b) with $r_m = 1.050$ to calculate

$$\mu_m = \alpha_0 + \frac{\alpha_1 \beta_0}{\alpha_0 - r_m} = 1.1930.$$

Similarly, for $r_m = 1.055$, $\mu_m = 1.1959$ so the change in the expected return on the market portfolio implied by the change in the risk free return is 0.0029.

(c) From Theorem 4.1(b) with $r_m = 1.05$ we calculate $t_m = 0.0701$. The implied market portfolio is now $x_m = h_0 + t_m h_1$, that is,

$$x_m = \begin{bmatrix} 0.3294 \\ 0.1353 \\ 0.1500 \\ 0.1928 \\ 0.1925 \end{bmatrix}.$$

◊

4.2 Optimization Derivation

We next analyze the problems addressed in Section 4.1 from a different point of view. Suppose r_m is already specified and we would like to find the point on the efficient frontier corresponding to the implied market portfolio. In Figure 4.2, we assume r_m is known and we would like to determine (σ_m, μ_m). One way to solve this is as follows. Let (σ_p, μ_p) represent an arbitrary portfolio and consider the slope of the

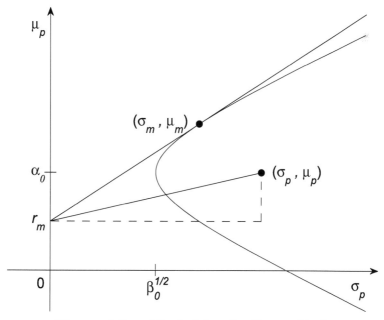

Figure 4.2   Maximizing the Sharpe Ratio.

line joining $(0, r_m)$ and (σ_p, μ_p). The slope of this line is $(\mu_p - r_m)/\sigma_p$ and is called the *Sharpe ratio*. In Figure 4.2 we can see the Sharpe ratio increases as (σ_p, μ_p) approaches (σ_m, μ_m). The market portfolio corresponds to where the Sharpe ratio is maximized. We thus need to solve the problem

$$\max \left\{ \frac{\mu'x - r_m}{(x'\Sigma x)^{\frac{1}{2}}} \mid l'x = 1 \right\}. \tag{4.6}$$

The problem (4.6) appears somewhat imposing, as the function to be maximized is the ratio of a linear function to the square root of a quadratic function. Furthermore, it appears from Figure 4.2 that in (4.6) we should be requiring that $\mu'x > r_m$ and we do this implicitly rather than making it an explicit constraint. The problem (4.6) can be analyzed using Theorem 1.5 as follows.

From Theorem 1.6(b), the gradient of the objective function for (4.6) is

$$\frac{(x'\Sigma x)^{\frac{1}{2}}\mu - (\mu'x - r_m)(x'\Sigma x)^{-\frac{1}{2}}\Sigma x}{x'\Sigma x}. \tag{4.7}$$

After rearranging, Theorem 1.5 states that the optimality conditions for (4.6) are

$$\left[\frac{x'\Sigma x}{\mu'x - r_m} \right]\mu - \Sigma x = \left[\frac{(x'\Sigma x)^{3/2}}{\mu'x - r_m} \right]u \, l, \quad l'x = 1, \tag{4.8}$$

where u is the multiplier for the budget constraint. Note that the quantities within square brackets are scalars, whereas μ, Σx and l are vectors.

Now, recall the optimization problem (2.3) and its optimality conditions (2.4). We repeat these for convenience:

$$\min\{ -t\mu'x + \frac{1}{2}x'\Sigma x \mid l'x = 1\} \tag{4.9}$$

$$t\mu - \Sigma x = ul, \text{ and, } l'x = 1. \tag{4.10}$$

Comparing (4.8) with (4.10) shows that the optimality conditions, and thus the optimal solutions for (4.9) and (4.6) will be identical provided

$$t = \frac{x'\Sigma x}{\mu'x - r_m}, \tag{4.11}$$

or equivalently

$$r_m = \mu'x - \frac{x'\Sigma x}{t}. \tag{4.12}$$

We have just shown the following:

Theorem 4.2 (a) For fixed r_m, let x_0 be optimal for (4.6). Then x_0 is also optimal for (4.9) with $t = x_0'\Sigma x_0/(\mu'x_0 - r_m)$.

(b) For fixed $t = t_1$, let $x_1 \equiv x(t_1)$ be optimal for (4.9). Then x_1 is optimal for (4.6) with $r_m = \mu'x_1 - x_1'\Sigma x_1/t_1$.

Theorems 4.1 (b) and 4.2 (a) are closely related. Indeed, Theorem 4.2 (a) implies the results of Theorem 4.1 (b) as we now show.

Let r_m be given and let x_0 be optimal for (4.6). Then Theorem 4.2 (a) asserts that x_0 is optimal for (2.3) with

$$t = x_0'\Sigma x_0/(\mu'x_0 - r_m). \tag{4.13}$$

Because x_0 is on the efficient frontier for (2.3) we can write $x_0'\Sigma x_0 = \beta_0 + \beta_2 t^2$ and $\mu'x_0 = \alpha_0 + \alpha_1 t$. Substitution of these two expressions in t given by Theorem 4.2 (a) gives

$$t = t_m = \frac{\beta_0 + \beta_2 t_m^2}{\alpha_0 + \alpha_1 t_m - r_m}.$$

Cross multiplying gives

$$t_m \alpha_0 + t_m^2 \alpha_1 - t_m r_m = \beta_0 + \beta_2 t_m^2.$$

Simplifying and solving for t_m now gives

$$t_m = \frac{\beta_0}{\alpha_0 - r_m},$$

which is the identical result obtained in Theorem 4.1 (b).

We can perform a similar analysis beginning with Theorem 4.2 (b). Set $t_m = t_1$ and $x_m = x_1$ in Theorem 4.2 (b). Because x_m is optimal for (2.3), $\mu_m = \alpha_0 + t_m \alpha_1$ and $\sigma_m^2 = \beta_0 + t_m^2 \beta_2$. Then Theorem 4.2(b) implies

$$
\begin{aligned}
r_m &= \mu' x_m - \frac{x_m' \Sigma x_m}{t_m} \quad\quad\quad\quad (4.14)\\
&= \alpha_0 + t_m \alpha_1 - \frac{\beta_0 + t_m^2 \beta_2}{t_m}.
\end{aligned}
$$

Multiplying through by t_m, using $\alpha_1 = \beta_2$ and simplifying gives

$$t_m = \left[\frac{\beta_0}{\alpha_0 - r_m} \right] \quad \text{and} \quad \mu_m = \alpha_0 + \left[\frac{\beta_0}{\alpha_0 - r_m} \right] \alpha_1, \quad\quad (4.15)$$

in agreement with Theorem 4.1(b).

There are other somewhat simpler methods to derive the results of (4.15) (see Exercise 4.4). Using Theorem 4.2(b) for the analysis has the advantage of being easy to generalize for practical portfolio optimization as we shall see in Section 9.1.

We next consider a variation of the previous problem. We assume that the value of the maximum Sharpe ratio, denoted by R, is known. R is identical to the slope of a CML. The questions to be answered are to what market portfolio does R correspond, and what is the associated implied risk free return r_m? The situation is illustrated in Figure 4.3.

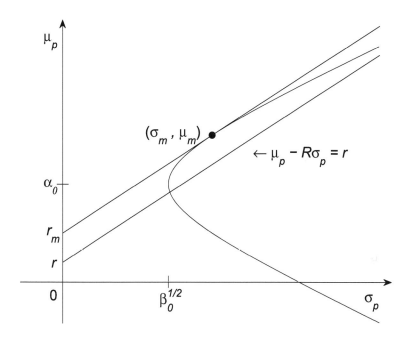

Figure 4.3   Maximize Implied Risk Free Return for Given Sharpe Ratio.

The line $\mu_p - R\sigma_p = r$ has slope R and is parallel to the specified CML. As r is increased (or equivalently $\mu_p - R\sigma_p$ is increased), the line will eventually be identical with the CML. The optimization problem to be solved is thus

$$\max\{\mu'x - R(x'\Sigma x)^{\frac{1}{2}} \mid l'x = 1\}. \qquad (4.16)$$

This can be rewritten as an equivalent minimization problem

$$\min\{-\mu'x + R(x'\Sigma x)^{\frac{1}{2}} \mid l'x = 1\}. \qquad (4.17)$$

The problem (4.17) is of the same algebraic form as (1.11) so that we can apply the optimality conditions of Theorem 1.5 to it. The dual feasibility part of the optimality conditions for (4.17) states

(see Exercise 4.5)

$$\mu - R\frac{\Sigma x}{(x'\Sigma x)^{\frac{1}{2}}} = ul, \tag{4.18}$$

where u is the multiplier for the budget constraint. This can be rewritten as

$$\left[\frac{(x'\Sigma x)^{\frac{1}{2}}}{R}\right]\mu - \Sigma x = \left[\frac{(x'\Sigma x)^{\frac{1}{2}}}{R}u\right]l. \tag{4.19}$$

Comparing (4.19) with (4.10) shows that the optimality conditions, and thus the optimal solutions for (4.9) and (4.19) will be identical provided

$$t = \frac{(x'\Sigma x)^{\frac{1}{2}}}{R},$$

or equivalently

$$R = \frac{(x'\Sigma x)^{\frac{1}{2}}}{t}. \tag{4.20}$$

Analogous to Theorem 4.2, we have

Theorem 4.3 (a) For fixed Sharpe ratio R, let x_0 be optimal for (4.17). Then x_0 is also optimal for (4.9) with $t = \frac{(x_0'\Sigma x_0)^{\frac{1}{2}}}{R}$.
(b) For fixed $t = t_1$, let $x_1 \equiv x(t_1)$ be optimal for (4.9). Then x_1 is optimal for (4.17) with $R = \frac{(x_1'\Sigma x_1)^{\frac{1}{2}}}{t_1}$.

Theorems 4.1(c) and 4.3(b) are closely related. Indeed, Theorem 4.3(b) implies the results of Theorem 4.1(c) as we now show. Let R be the given slope of the CML, assume

$$R > \sqrt{\beta_2} \tag{4.21}$$

and let $t_m = t_1$ in the statement of Theorem 4.3(b). We can determine t_m and r_m as follows. Theorem 4.3(b) asserts that

$$R = \frac{(x_m'\Sigma x_m)^{1/2}}{t_m}.$$

Squaring and cross multiplying give

$$t_m^2 R^2 = \beta_0 + t_m^2 \beta_2.$$

Solving this last for t_m gives

$$t_m = \frac{\beta_0^{1/2}}{(R^2 - \beta_2)^{1/2}}. \qquad (4.22)$$

In addition, we can determine the implied risk free return as follows. The CML is $\mu_p = r_m + R\sigma_p$. The point (σ_m, μ_m) is on this line, so $\mu_m = r_m + R\sigma_m$. Therefore,

$$
\begin{aligned}
\alpha_0 + t_m \alpha_1 &= r_m + R(\beta_0 + t_m^2 \beta_2)^{1/2} \\
&= r_m + R \left[\beta_0 + \frac{\beta_0 \beta_2}{R^2 - \beta_2} \right]^{1/2} \\
&= r_m + R \left[\frac{\beta_0 R^2 - \beta_0 \beta_2 + \beta_0 \beta_2}{R^2 - \beta_2} \right]^{1/2} \\
&= r_m + R^2 \left[\frac{\beta_0}{R^2 - \beta_2} \right]^{1/2} \\
&= r_m + R^2 t_m.
\end{aligned}
$$

Solving for r_m in this last gives

$$
\begin{aligned}
r_m &= \alpha_0 + t_m(\alpha_1 - R^2) \\
&= \alpha_0 - \frac{\beta_0^{1/2}}{(R^2 - \beta_2)^{1/2}}(R^2 - \beta_2) \\
&= \alpha_0 - [\beta_0(R^2 - \beta_2)]^{1/2}. \qquad (4.23)
\end{aligned}
$$

Equations (4.23) and (4.22) show that Theorem 4.1(c) is a special case of Theorem 4.3(b).

There are other somewhat simpler methods to deduce the previous result (see Exercise 4.6). The advantage of using Theorem 4.3(b) is that it generalizes in a straightforward manner to practical portfolio optimization problems, as we shall see in Sections 9.1.

Implicit in (4.23) is the assumption that $R^2 > \beta_2$. To see why this is reasonable, observe that from (4.3) the slope of the efficient frontier is

$$\frac{d\mu_p}{d\sigma_p} = \frac{\alpha_1 \sigma_p}{(\alpha_1(\sigma_p^2 - \beta_0))^{\frac{1}{2}}}$$

$$= \frac{\beta_2^{\frac{1}{2}}}{(1 - \frac{\beta_0}{\sigma_p^2})^{\frac{1}{2}}}. \qquad (4.24)$$

From (4.24) it follows that

$$\frac{d\mu_p}{d\sigma_p} > \beta_2^{\frac{1}{2}},$$

and

$$\lim_{\sigma_p \to \infty} \frac{d\mu_p}{d\sigma_p} = \beta_2^{\frac{1}{2}}.$$

We complete this section by stating a theorem which summarizes Theorems 4.2(b) and 4.3(b).

Theorem 4.4 *Let* $x(t_1) = x_1$ *be a point on the efficient frontier for* $t = t_1$. *Then* $(\mu' x_1 - r_m)/(x_1' \Sigma x_1)^{\frac{1}{2}}$ *is the maximum Sharpe ratio for* $r_m = \mu' x_1 - x_1' \Sigma x_1 / t_1$, *and,* $r_m = \mu' x_1 - R(x_1' \Sigma x_1)^{\frac{1}{2}}$ *is the maximum risk free return with Sharpe ratio* $R = (x_1' \Sigma x_1)^{\frac{1}{2}} / t_1$.

4.3   Free Problem Solutions

Suppose we have solved

$$\min\{-t\mu' x + \frac{1}{2} x' \Sigma x \mid l' x = 1\}$$

and obtained h_0, h_1, α_0, $\alpha_1 = \beta_2$ and β_0. If we pick any $t_0 > 0$ then we have a point on the efficient frontier and we can compute $\mu_{p0} = \alpha_0 + t_0 \alpha_1$

and $\sigma_{p0}^2 = \beta_0 + t_0^2\beta_2$. But then (1) μ_{p0} is the optimal solution for the problem of maximizing the portfolio's expected return with the portfolio variance being held at σ_{p0}^2; i.e., (2.2) with $\sigma_p^2 = \sigma_{p0}^2$. Furthermore, (2) σ_{p0}^2 is the minimum variance of the portfolio with the expected return being held at μ_{p0}; i.e., (2.1) with $\mu_p = \mu_{p0}$.

Table 4.1 shows these pairs for a variety of values of t for the problem of Example 2.1. For example, for $t = 0.10$, the minimum portfolio variance is 0.0134043 when the expected return is required to be 1.19574. Conversely, the maximum expected return is 1.19574 when the variance is held at 0.0134043.

TABLE 4.1   Minimum Variance and Maximum Expected Returns For Example 2.1.

t	Minimum Variance	For Specified Return μ_p	Maximum Expected Return	For Specified Variance σ_p^2
0.1000	0.0134043	1.19574	1.19574	0.0134043
0.1111	0.0148017	1.20236	1.20236	0.0148017
0.1250	0.0167553	1.21064	1.21064	0.0167553
0.1429	0.0196049	1.22128	1.22128	0.0196049
0.1667	0.0239953	1.23546	1.23546	0.0239953
0.2000	0.0312766	1.25532	1.25532	0.0312766
0.2500	0.0446809	1.28511	1.28511	0.0446809
0.3333	0.0736407	1.33475	1.33475	0.0736407
0.5000	0.156383	1.43404	1.43404	0.156383
1.0000	0.603191	1.73191	1.73191	0.603191

Theorem 4.4 also says we have solved two other problems at t_0. These are; (3)

$$(\mu_{p0} - r_m)/(\sigma_{p0}^2)^{1/2} = (\mu_{p0} - r_m)/\sigma_{p0}^2$$

is the maximum Sharpe ratio for $r_m = \mu_{p0} - \sigma_{p0}^2/t_0$, and, (4) $r_m = \mu_{p0} - R(\sigma_{p0}^2)^{1/2}$ is the maximum risk free return with Sharpe ratio $R = (\sigma_{p0}^2)^{1/2}/t_0$.

Table 4.2 shows these pairs for a variety of values of t for the problem of Example 2.1. For example, for $t = 0.10$, the maximum Sharpe ratio is 1.15777 for a fixed risk free return of 1.0617. Conversely, the maximum risk free return is 1.0617 for a specified Sharpe ratio of 1.15777.

TABLE 4.2   Maximum Sharpe Ratios and Maximum Risk Free Returns For Example 2.1.

t	Maximum Sharpe ratio	For Risk Free Return r_m	Maximum Risk Free Return r_m	For Specified Sharpe Ratio R
0.1000	1.15777	1.0617	1.0617	1.15777
0.1111	1.09496	1.06915	1.06915	1.09496
0.1250	1.03554	1.0766	1.0766	1.03554
0.1429	0.980122	1.08404	1.08404	0.980122
0.1667	0.929424	1.09149	1.09149	0.929424
0.2000	0.88426	1.09894	1.09894	0.88426
0.2500	0.845514	1.10638	1.10638	0.845514
0.3333	0.814104	1.11383	1.11383	0.814104
0.5000	0.790906	1.12128	1.12128	0.790906
1.0000	0.776654	1.12872	1.12872	0.776654

Note that in Table 4.2 the two expressions for Sharpe ratios, namely $(\mu_{p0} - r_m)/(\sigma_{p0}^2)^{1/2}$ and $R = (\sigma_{p0}^2)^{1/2}/t_0$ are numerically identical. Furthermore, the two expressions for the risk free return, namely $r_m = \mu_{p0} - \sigma_{p0}^2/t_0$ and $r_m = \mu_{p0} - R(\sigma_{p0}^2)^{1/2}$ produce identical numerical results. The equality of these two pairs of expressions is true in general. See Exercise 4.8.

4.4   Computer Programs

Figure 4.4 displays the routine Example4p1.m which performs the calculations for Example 4.1. The data is defined in lines 1-6 and line 8 invokes the routine EFMVcoeff (see Figure 2.6) to compute the coefficients for the efficient frontier. The computations for parts (a), (b) and (c) of Example 4.1 are then done in a straightforward manner.

```
1   %Example4p1.m
2   mu = [ 1.1 1.15 1.2 1.27 1.3 ]'
3   Sigma = [1.e-2 0.7e-3 0 0 0.6e-3 ;
4        0.7e-3 5.0e-2 1.e-4 0 0 ;
5        0 1.e-4  7.0e-2 0 0 ;
6        0 0 0 8e-2 0 ;
7        0.6e-3 0 0 0 9.e-2]
8   checkdata(Sigma,1.e-6);
9   [alpha0,alpha1,beta0,beta2,h0,h1] = EFMVcoeff(mu,Sigma);
10
11  %Example 4.1(a)
12  string = 'Example 4.1(a)'
13  mum = 1.4
14  rm = alpha0 - (alpha1*beta0)/(mum - alpha0)
15  tm = beta0/(alpha0-rm)
16  mum = alpha0 + tm * alpha1
17  sigp2 = beta0 + tm*tm*beta2
18
19  %Example 4.1(b)
20  string = 'Example 4.1(b)'
21  rm = 1.05000
22  tm = beta0/(alpha0 - rm)
23  mum = alpha0 + alpha1*tm
24  sigp2 = beta0 + tm*tm*beta2
25
26  rm = 1.055
27  tm = beta0/(alpha0 - rm)
28  mum = alpha0 + alpha1*tm
29  sigp2 = beta0 + tm*tm*beta2
30
31  %Example 4.1(c)
32  string = 'Example 4.1(c)'
33  rm = 1.05
34  tm = beta0/(alpha0 - rm)
35  port = h0 + tm * h1
```

Figure 4.4   Example4p1.m.

```
 1  %SharpeRatios.m
 2  mu = [ 1.1 1.2 1.3 ]'
 3  Sigma = [1.e−2 0 0 ; 0 5.0e−2 0 ; 0 0 7.0e−2 ]
 4  checkdata(Sigma,1.e−6);
 5  [alpha0,alpha1,beta0,beta2,h0,h1] = EFMVcoeff(mu,Sigma);
 6  for i = 1:10
 7  t = 1/(11−i)
 8  x1 = h0 + t.*h1;
 9  mup = mu'*x1;
10  sigp2 = x1' * Sigma * x1;
11  str = 'Minimum variance %g for fixed expected return of
12  %g\n';
13  fprintf(str,sigp2,mup)
14  str = 'Maximum expected return %g for fixed variance of
15  %g\n';
16  fprintf(str,mup,sigp2)
17  rm = mup − sigp2/t;
18  Sratio = (mup − rm)/sigp2^(1/2);
19  fprintf('Maximum Sharpe ratio is %g for rm %g\n',
20  Sratio,rm)
21  Sratio = sigp2^(1./2.)/t;
22  rm =  mup − Sratio*sigp2^(1/2);
23  fprintf('Max risk free return is %g for Sharpe ratio
24  %g\n'...
25  ,rm,Sratio)
26  end
```

Figure 4.5   SharpeRatios.m.

Figure 4.5 shows the routine SharpeRatios.m which computes the solution for four problems at each of ten points on the efficient frontier for the data of Example 2.1. Lines 2 and 3 define the data and line 4 checks the covariance matrix. Line 5 uses the function EFMVcoeff (see Figure 2.6) to calculate the coefficients of the efficient frontier. The "for" loop from line 6 to 20, constructs various values of t. For each such t, line 8 constructs the efficient portfolio for t and lines 9 and 10 compute the mean and variance of that portfolio. Then, as discussed in Section 4.3, the computed portfolio mean is the maximum portfolio mean with the variance being held at the computed level. Conversely, the computed variance is the minimum variance when the portfolio expected return is held at the computed level. These values are summarized in Table 4.1.

4.5 Exercises

4.1 For the data of Example 2.1, determine the implied risk free return corresponding to (a) $\mu_p = 1.2$, (b) $\mu_p = 1.25$.

4.2 What market portfolio would give an implied risk free return of $r_m = 0$?

4.3 Suppose $r_m = 1.1$ is the risk free return for the data of Example 2.1. Find the implied market portfolio.

4.4 If we use the fact that the optimal solution must lie on the efficient frontier, then (4.6) can be formulated as

$$\max \left\{ \frac{\alpha_0 + \alpha_1 t - r_m}{(\beta_0 + \beta_2 t^2)^{\frac{1}{2}}} \right\},$$

where the maximization is over all $t \geq 0$. Show the optimal solution for this is identical to (4.15).

4.5 Verify (equation 4.18).

4.6 If we use the fact that the optimal solution must lie on the efficient frontier, then (4.17) can be formulated as

$$\max\{\alpha_0 + t\alpha_1 - R(\beta_0 + \beta_2 t^2)^{\frac{1}{2}}\}, \tag{4.25}$$

where the maximization is over all $t \geq 0$. Show that the optimal solution for this is identical to (4.22).

4.7 Given the coefficients of the efficient frontier $\alpha_0 = 1.13$, $\alpha_1 = 0.72$, $\beta_0 = 0.007$ and $\beta_2 = 0.72$.

    (a) Find the risk free return r_m implied by a market expected return of $\mu_m = 1.37$. Also find the associated t_m, μ_m and σ_m^2. What market expected return is implied by r_m?

(b) Suppose the expected return on the market changes from $\mu_m = 1.35$ to $\mu_m = 1.36$. What is the implied change in the risk free return?

4.8 Let μ_{p0} and σ_{p0} be as in Section 4.3.

(a) Show the Sharpe ratios $(\mu_{p0} - r_m)/(\sigma_{p0}^2)^{1/2}$ and $R = (\sigma_{p0}^2)^{1/2}/t_0$ are identical, where $r_m = \mu_{p0} - \sigma_{p0}^2/t_0$.

(b) Show the risk free returns $r_m = \mu_{p0} - \sigma_{p0}^2/t_0$ and $r_m = \mu_{p0} - R(\sigma_{p0}^2)^{1/2}$ are identical, where $R = (\sigma_{p0}^2)^{1/2}/t_0$.

Chapter 5

Quadratic Programming Geometry

This chapter develops the essential ideas of quadratic programming (QP). Section 5.1 shows how to solve a 2-dimensional QP by geometrical reasoning. In Section 5.2, optimality conditions are formulated for a QP, again using geometrical reasoning. In Section 5.3, it is shown how to draw elliptic level sets for a quadratic function. In Section 5.4, optimality conditions for a QP having n variables and m inequality constraints, are formulated and proven to be sufficient for optimality.

5.1   The Geometry of QPs

The purpose of this section is to introduce the most important properties of quadratic programming by means of geometrical examples. We first show how to determine an optimal solution geometrically. A major difference between linear and quadratic programming problems is the number of constraints active at an optimal solution and this is illustrated by means of several geometrical examples.

We begin our analysis of a quadratic programming problem by observing properties of an optimal solution in several examples.

Example 5.1

$$
\begin{aligned}
\text{minimize:} \quad & -4x_1 - 10x_2 + x_1^2 + x_2^2 \\
\text{subject to:} \quad -x_1 + x_2 &\leq 2, && (1) \\
x_1 + x_2 &\leq 6, && (2) \\
x_1 &\leq 5, && (3) \\
-x_2 &\leq 0, && (4) \\
-x_1 &\leq -1. && (5)
\end{aligned}
$$

The *objective function* for this problem is $f(x) = -4x_1 - 10x_2 + x_1^2 + x_2^2$. The *feasible region*, denoted by R, is the set of points

$$
x = \begin{bmatrix} x_1 \\ x_2 \end{bmatrix}
$$

which simultaneously satisfy constraints (1) to (5). An *optimal solution* is a feasible point for which the objective function is smallest among those points in R. The feasible region has the same form as that for a linear programming problem. The difference between a quadratic and a linear programming problem is that the former has a quadratic objective function while the latter has a linear objective function. A linear programming problem is thus a special case of a quadratic programming problem in which the coefficients of all the quadratic terms in the objective function are zero.

Because this example has only two variables, we can obtain a graphical solution for it. The feasible region is shown in Figure 5.1. The behavior of the objective function can be determined by observing that the set of points for which $f(x) = $ a constant, is a circle centered at $(2, 5)'$ and that points inside the circle give a smaller value of f. Figure 5.1 shows the circle $f(x) = -25$. Each such circle (more generally, an ellipse) is called a *level set* of f. Points inside this circle and in R, shown in the shaded region of Figure 5.1, give a better objective

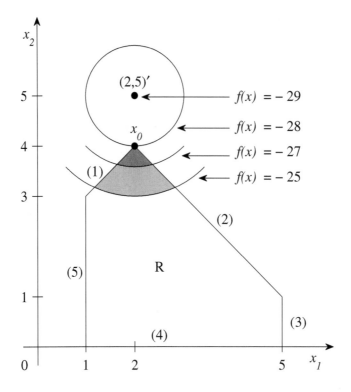

Figure 5.1   Geometry of Example 5.1.

function value. Also shown is the analogous situation for $f(x) = -27$. Points in the heavily shaded area are more attractive than those on the circle segment $f(x) = -27$. Intuitively, we can obtain an optimal solution by decreasing the radius of the circle until it intersects R at a single point. Doing so gives the optimal solution $x_0 = (2, 4)'$ with $f(x_0) = -28$.

Note that the optimal solution for Example 5.1 is an *extreme point* (geometrically a corner of the feasible region) just as in linear programming. This is not always the case, however, as we show in the following two examples.

Example 5.2

$$\text{minimize:} \quad -10x_1 - 6x_2 + x_1^2 + x_2^2$$

$$\text{subject to:} \quad -\ x_1 + x_2 \leq 2, \tag{1}$$

$$x_1 + x_2 \leq 6, \tag{2}$$

$$x_1 \qquad \leq 5, \tag{3}$$

$$-\ x_2 \leq 0, \tag{4}$$

$$-\ x_1 \qquad \leq -1. \tag{5}$$

The feasible region for this example is identical to that of Example 5.1. The linear terms in the new objective function $f(x) = -10x_1 - 6x_2 + x_1^2 + x_2^2$ have been changed so that the level sets are concentric circles centered at $(5, 3)'$. Figure 5.2 shows the feasible region and level

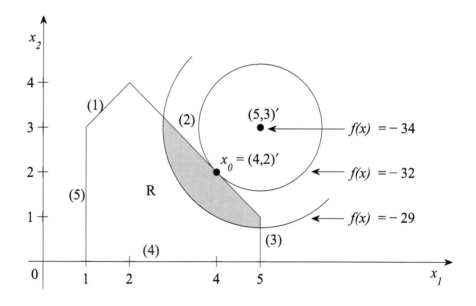

Figure 5.2   Geometry of Example 5.2.

set $f(x) = -29$. Points inside this level set and in R give a better objective function value. "Shrinking" the level sets as in Example 5.1 gives the optimal solution $x_0 = (4, 2)'$.

Note that at the optimal solution for Example 5.2 the active constraint is tangent to the level set of f. Also observe that x_0 is **not** an extreme point: only one constraint [namely constraint (2)] is active at x_0. It is also possible for no constraints to be active at the optimal solution as shown in our third example.

Example 5.3

$$
\begin{aligned}
\text{minimize:} \quad & -6x_1 - 4x_2 + x_1^2 + x_2^2 \\
\text{subject to:} \quad & -x_1 + x_2 \leq 2, \quad & (1) \\
& x_1 + x_2 \leq 6, \quad & (2) \\
& x_1 \leq 5, \quad & (3) \\
& -x_2 \leq 0, \quad & (4) \\
& -x_1 \leq -1. \quad & (5)
\end{aligned}
$$

The constraints are identical to those of the previous two examples. The level sets of the new objective function are circles centered at $(3, 2)'$. Figure 5.3 shows the level set for $f(x) = -12$. Since the center of this level set lies within R, it follows that the optimal solution x_0 is precisely this point.

Based on Examples 5.1, 5.2, and 5.3, we can make the following observation.

Observation 1.

Any number of constraints can be active at the optimal solution of a quadratic programming problem.

It is this observation that distinguishes a quadratic programming problem from a linear one. Although the algorithms for the solution of a quadratic programming problem are quite similar to those for linear programming problems, the principal difference is the method of accounting for a varying number of active constraints.

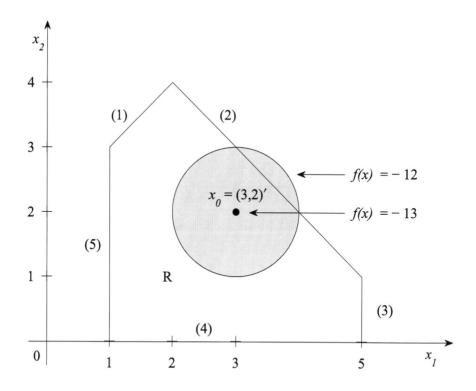

Figure 5.3   Geometry of Example 5.3.

5.2   Geometry of QP Optimality Conditions

An optimal solution for a linear programming problem is characterized by the condition that the negative gradient of the objective function lies in the cone spanned by the gradients of those constraints active at it. Under a simple assumption, the same characterization holds for a quadratic programming problem. Recalling that the *gradient* of a function is just the vector of first partial derivatives, the gradient of the objective functions for the previous three examples is

$$\nabla f(x) \;\equiv\; g(x) \;\equiv\; \left[\begin{array}{c} \dfrac{\partial f(x)}{\partial x_1} \\[2ex] \dfrac{\partial f(x)}{\partial x_2} \end{array}\right].$$

For brevity, we will use $g(x) = \nabla f(x)$ to denote the gradient of f at x. Geometrically, the gradient of a function at a point x_0 is orthogonal to the tangent of the level set at x_0.

Figure 5.4 shows the negative gradient of the objective function for Example 5.1 at several points; x_0, x_1, x_2, and x_3. Note that in Figure 5.4, we have used y_1 and y_2 to denote the coordinate axes, and x_0, x_1, x_2, and x_3 to denote points in the plane.

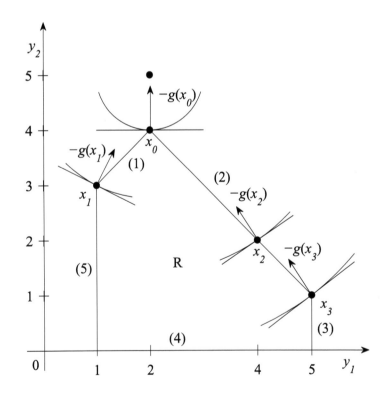

Figure 5.4   Gradient vectors for objective function of Example 5.1.

The negative gradient is drawn at x_0, for example, as follows. The level set at x_0 is first drawn. This is the circle centered at $(2,5)'$ which passes through x_0. Next, the tangent to this level set is drawn at x_0. Since $-g(x_0)$ is orthogonal to this tangent, it must point either straight up or straight down. The gradient of a function points in the direction of maximum local increase of the function, so $-g(x_0)$ must point in the direction of maximum local decrease; that is, toward the center of the circle. The remaining negative gradients are constructed using a similar procedure.

Let a_i, $i = 1, 2, \ldots, 5$, denote the gradient of the i-th constraint function for Example 5.1. Then

$$a_1 = \begin{bmatrix} -1 \\ 1 \end{bmatrix}, \quad a_2 = \begin{bmatrix} 1 \\ 1 \end{bmatrix}, \ldots, a_5 = \begin{bmatrix} -1 \\ 0 \end{bmatrix}.$$

The a_i's can be drawn using the same procedure as for $g(x)$. Because the constraint functions are linear, their gradients are constant, whereas the objective function is quadratic and therefore its gradient is linear. The gradients of the constraints are orthogonal to the boundaries of R, and point out and away from R.

Figure 5.5 shows $-g(x)$ and the cone spanned by the gradients of the constraints active at x for each of the points $x = x_0$, x_1, x_2, and x_3. From this, we observe that at the optimal solution x_0, $-g(x_0)$ does indeed lie in the cone spanned by the gradients of the active constraints a_1 and a_2. Furthermore, that condition is **not** satisfied at any of the nonoptimal points x_1, x_2, and x_3. At x_1, for example, there are two active constraints, namely (1) and (5), and $-g(x_1)$ lies outside the cone spanned by a_1 and a_5. At x_2, only one constraint is active [constraint (2)] and the cone spanned by a_2 is the set of vectors pointing in the same direction as a_2. Since $-g(x_2)$ is not parallel to a_2, it is not in this cone.

Observation 2.

A point is optimal if and only if it is feasible and the negative gradient of the objective function at the point lies in the cone spanned by the gradients of the active constraints.[1]

---

[1]We will show (Theorem 5.1) that this statement is valid generally, provided the objective function is a convex function (see Section 5.4).

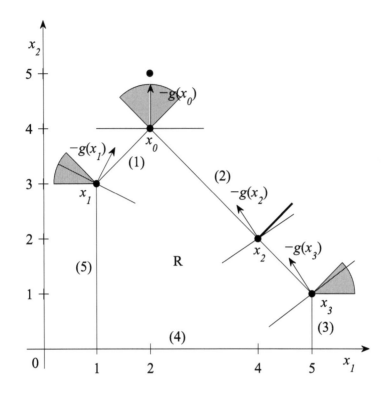

Figure 5.5   Optimality conditions for Example 5.1.

We have given a geometrical verification that for Example 5.1 x_0 satisfies the optimality conditions formulated in Observation 2. We next do the same thing algebraically. For this example we have

$$g(x) = \begin{bmatrix} -4 + 2x_1 \\ -10 + 2x_2 \end{bmatrix},$$

so that $g(x_0) = (0, -2)'$. The cone part of the optimality conditions requires that there are nonnegative numbers u_1 and u_2 satisfying

$$-g(x_0) = u_1 a_1 + u_2 a_2 .$$

These are precisely the 2 linear equations in 2 unknowns

$$- u_1 + u_2 = 0,$$
$$u_1 + u_2 = 2.$$

These have solution $u_1 = u_2 = 1$. Since both u_1 and u_2 are nonnegative, we have algebraically verified that the optimality conditions are satisfied.

The optimality conditions for Example 5.2 can be analyzed in a similar manner. Recall from Figure 5.2 that the optimal solution is $x_0 = (4, 2)'$ and that only constraint (2) is active at x_0. The cone spanned by the gradient of this single constraint is precisely the set of positive multipliers of a_2. From Figure 5.6, it is clear that $-g(x_0)$ points in the same direction as a_2 so that the optimality conditions are indeed satisfied at x_0.

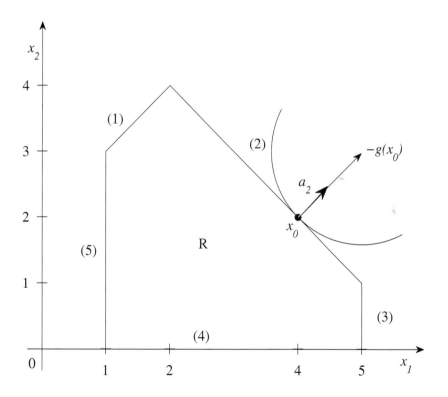

Figure 5.6   Optimality conditions for Example 5.2.

To provide algebraic verification of the optimality conditions for Example 5.2, we observe that

$$g(x) = \begin{bmatrix} -10 + 2x_1 \\ -6 + 2x_2 \end{bmatrix}$$

so that $g(x_0) = (-2, -2)'$. The cone part of the optimality conditions requires that there is a nonnegative number u_2 with

$$-g(x_0) = u_2 a_2 .$$

These are precisely the 2 linear equations in 1 unknown

$$u_2 = 2,$$
$$u_2 = 2,$$

which have the obvious solution $u_2 = 2$. Since u_2 is nonnegative, we have algebraically verified that the optimality conditions are satisfied.

Finally, the optimal solution $x_0 = (3, 2)'$ for Example 5.3 lies in the interior of R. Since no constraints are active at x_0, the cone part of the optimality conditions requires that $-g(x_0) = 0$. Since

$$g(x) = \begin{bmatrix} -6 + 2x_1 \\ -4 + 2x_2 \end{bmatrix},$$

it follows that $-g(x_0) = 0$ and we have algebraically verified that the optimality conditions are satisfied.

These three examples are the basis for Observations 1 and 2. The number of active constraints at the optimal solution for Examples 5.1, 5.2, and 5.3 is 2, 1, and 0, respectively. In each example, the optimal solution is characterized by the condition that it is feasible and that the gradient of the objective function lies in the cone spanned by the gradients of the active constraints.

5.3 The Geometry of Quadratic Functions

The objective functions for Examples 5.1, 5.2, and 5.3 all have the two properties: that the coefficients of x_1^2 and x_2^2 are identical, and that there are no terms involving $x_1 x_2$. Consequently, the level sets for each objective function are circles which can be drawn by inspection. However, for the function

$$f(x) = -67x_1 - 59x_2 + \tfrac{13}{2}x_1^2 + 5x_1 x_2 + \tfrac{13}{2}x_2^2,$$

it is not immediately obvious how the level sets should be drawn. The purpose of this section is to show a method by which the level sets can easily be drawn. The relevant mathematical tools are eigenvectors and eigenvalues. The reader unfamiliar with these concepts may wish to consult an appropriate reference (Noble, 1969, for example). However, it is our intention that this section be reasonably self-contained.

Consider a general quadratic function of two variables

$$f(x) = c'x + \tfrac{1}{2}x'Cx, \tag{5.1}$$

where x and c are 2-vectors and C is a $(2,2)$ symmetric matrix. Let S denote the $(2,2)$ matrix whose columns are the eigenvectors of C and let D denote the $(2,2)$ diagonal matrix of corresponding eigenvalues. The defining property of S and D is

$$CS = SD. \tag{5.2}$$

An elementary property of eigenvectors is that they are orthogonal. Provided that they are also scaled to have unit norm,

$$S'S = I, \tag{5.3}$$

where I denotes the $(2,2)$ identity matrix. Multiplying (5.2) on the left by S' and using (5.3) gives

$$S'CS = D. \tag{5.4}$$

Let x_0 be the point where $f(x)$ is minimized. Since this is an uncon-strained quadratic minimization problem, the optimality conditions of the previous section imply that the gradient of f at x_0 must be equal to zero. That is, x_0 is the solution of the linear equations

$$Cx_0 = -c.$$

We next introduce a change of variable y related to x by

$$x = Sy + x_0. \tag{5.5}$$

Taylor's series now gives

$$f(x) = f(x_0) + g(x_0)'Sy + \tfrac{1}{2}y'S'CSy. \tag{5.6}$$

We have chosen x_0 such that $g(x_0) = 0$. With this and (5.4), (5.6) simplifies to

$$f(x) = f(x_0) + \tfrac{1}{2}y'Dy. \tag{5.7}$$

We are now ready to make the key point. Because D is diagonal, (5.7) expresses f in y coordinates solely in terms of y_1^2 and y_2^2. In particular, (5.7) includes no linear terms in y_1 y_2, and no cross product terms involving y_1y_2. Using (5.7), it is easy to draw level sets for f in y coordinates.

Summarizing, level sets for f may be drawn by utilizing the follow-ing steps.

1. Obtain S and D by solving the eigenvalue problem for C.

2. Solve $Cx_0 = -c$.

3. Draw the y_1, y_2 axes in the x_1, x_2 space by drawing the column vectors of S centered at x_0.

4. Draw the level sets of f in y_1, y_2 space using (5.7).

The steps of the procedure are illustrated in the following example.

Example 5.4 Draw the level sets for

$$f(x) = -67x_1 - 59x_2 + \tfrac{13}{2}x_1^2 + 5x_1x_2 + \tfrac{13}{2}x_2^2.$$

Here

$$c = \begin{bmatrix} -67 \\ -59 \end{bmatrix}, \quad \text{and} \quad C = \begin{bmatrix} 13 & 5 \\ 5 & 13 \end{bmatrix}.$$

By definition, an eigenvector $s = (s_1, s_2)'$ of C and its corresponding eigenvalue λ satisfy

$$Cs = \lambda s, \quad \text{or} \quad (C - \lambda I)s = 0.$$

This last is a system of 2 linear equations in 2 unknowns. Because the right-hand side is zero, a necessary condition for the system to have nontrivial solutions is $\det(C - \lambda I) = 0$; that is

$$(13 - \lambda)^2 - 25 = 0.$$

This has roots $\lambda = 18$ and $\lambda = 8$ so

$$D = \begin{bmatrix} 18 & 0 \\ 0 & 8 \end{bmatrix}.$$

The eigenvector corresponding to $\lambda = 18$ is obtained by substituting into $(C - \lambda I)s = 0$. This gives $s_1 = s_2$ so that

$$s = s_2 \begin{bmatrix} 1 \\ 1 \end{bmatrix}.$$

Because (5.3) requires $s's = 1$, we choose $s_2 = 1/\sqrt{2}$ giving

$$s = \frac{1}{\sqrt{2}} \begin{bmatrix} 1 \\ 1 \end{bmatrix}.$$

Proceeding in a similar manner with $\lambda = 8$ gives the second eigenvector

$$s = \frac{1}{\sqrt{2}} \begin{bmatrix} -1 \\ 1 \end{bmatrix}.$$

Thus

$$S = \frac{1}{\sqrt{2}} \begin{bmatrix} 1 & -1 \\ 1 & 1 \end{bmatrix},$$

and this completes Step 1.

For Step 2, we solve the linear equations

$$\begin{bmatrix} 13 & 5 \\ 5 & 13 \end{bmatrix} x_0 = \begin{bmatrix} 67 \\ 59 \end{bmatrix}.$$

These have solution $x_0 = (4, 3)'$.

For Step 3, the origin of the y coordinate system is at x_0 in the x coordinate system. The y_1 axis extends from x_0 in the direction $2^{-1/2}(1\ ,\ 1)'$ and the y_2 axis extends from x_0 in the direction $2^{-1/2}(-1\ ,\ 1)'$. This is shown in Figure 5.7.

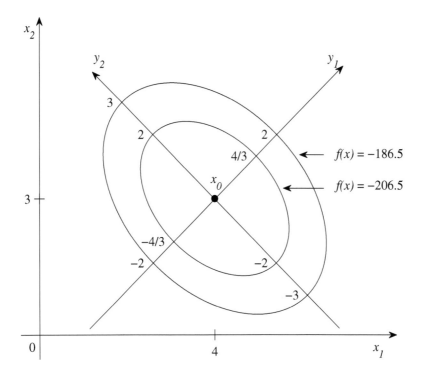

Figure 5.7   Level sets for Example 5.4.

For Step 4, we observe that from (5.7) with $f(x_0) = -222.5$, the level sets of f in y coordinates are

$$f(x) = -222.5 + 9y_1^2 + 4y_2^2 .$$

The level set for $f(x) = -186.5$ is thus

$$\frac{y_1^2}{4} + \frac{y_2^2}{9} = 1 \; .$$

This is an ellipse with intercepts at $(\pm 2, 0)'$ and $(0, \pm 3)'$, and is shown in Figure 5.7. Similarly, the level set for $f(x) = -206.5$ is

$$\frac{y_1^2}{4} + \frac{y_2^2}{9} = \frac{4}{9} \; ,$$

which is "parallel" to the previous one (see Figure 5.7).

5.4   Optimality Conditions for QPs

In Section 5.2, we saw for some two-dimensional examples that an optimal solution was characterized as being a feasible point for which the negative gradient of the objective function lies in the cone spanned by the gradients of the active constraints. In this section, we generalize this condition to problems having n variables and m linear inequality constraints.

We now consider a general convex[2] quadratic programming problem (QP);

$$\min\{c'x + \frac{1}{2}x'Cx \mid Ax \leq b\}, \tag{5.8}$$

where c, x are $n-$vectors, C is an (n, n) symmetric positive semidefinite matrix, A is an (m, n) matrix and b is an $m-$vector. The given problem thus has n variables (the components of x) and m linear inequality constraints. c, C, A, and b are to be considered given data and the goal of (5.8) is to determine the optimal value of x. It is sometimes useful

---

[2]In the present context, *convex* is equivalent to the Hessian of the objective function, C, being positive semidefinite.

to reformulate (5.8) so that the gradients of the constraints are made explicit:

$$\min\{c'x + \frac{1}{2}x'Cx \mid a_i'x \le b_i, \ i = 1, 2, \ldots, m\}, \tag{5.9}$$

where $a_1, a_2, \ldots, a_m$ are n−vectors, a_i is the gradient of the i-th constraint function,

$$A' = [a_1, a_2, \ldots, a_m] \text{ and } b = (b_1, b_2, \ldots, b_m)'.$$

We need to formalize a few concepts in definitions.

Definition 5.1 *The* **feasible region** *for (5.8) is the set of points* $R = \{x \mid Ax \le b\}$. *A point x_0 is* **feasible** *for (5.8) if $x_0 \in R$ and* **infeasible**, *otherwise. Constraint i is* **inactive** *at x_0 if $a_i'x_0 < b_i$,* **active** *at x_0 if $a_i'x_0 = b_i$ and* **violated** *at x_0 if $a_i'x_0 > b_i$.*

Definition 5.2 *Let $f(x) = c'x + \frac{1}{2}x'Cx$ denote the objective function for (5.8). A point x_0 is* **optimal** *for (5.8) if x_0 is feasible for (5.8) and $f(x_0) \le f(x)$ for all $x \in R$.*

We are now ready to formulate optimality conditions for (5.9). Let x_0 be an optimal solution for (5.9). Then by Definition 5.2, $Ax_0 \le b$. Furthermore, Examples 5.1, 5.2 and 5.3 suggest that $-\nabla f(x_0)$ must be a nonnegative linear combination of the gradients of those constraints which are active at x_0. We could write this in a somewhat cumbersome way as

$$-\nabla f(x_0) = \sum_{i:a_i'x_0=b_i} u_i a_i$$

and

$$u_i \ge 0, \text{ for all } i \text{ such that } a_i'x_0 = b_i.$$

There is a simpler way to formulate this last. Suppose we associate a multiplier u_i with *every* constraint i and then insist that $u_i = 0$ for every inactive constraint i. A convenient way to do this is by the two requirements

$$-\nabla f(x_0) = \sum_{i=1}^{m} u_i a_i, \quad u_i \geq 0, \quad i = 1, 2, \ldots, m \tag{5.10}$$

and

$$u_i(a_i' x_0 - b_i) = 0, \quad i = 1, 2, \ldots, m. \tag{5.11}$$

The requirement (5.11) forces u_i to be zero for each inactive constraint i and thus (5.10) does indeed require that the negative gradient of f at x_0 be a nonnegative linear combination of the gradients of those constraints active at x_0.

We summarize this in the following definition.

Definition 5.3 x_0 *satisfies the* **optimality conditions** *for (5.9) if*

$$a_i' x_0 \leq b_i, \ i = 1, 2, \ldots, m, \quad (primal \ feasibility)$$

$$-\nabla f(x_0) = \sum_{i=1}^{m} u_i a_i, \ u_i \geq 0, \ i = 1, 2, \ldots, m, \quad (dual \ feasibility)$$

$$u_i(a_i' x_0 - b_i) = 0, \ i = 1, 2, \ldots, m. \quad (complementary \\ slackness)$$

Definition 5.3 can be reformulated somewhat more compactly using matrix notation. We assert that x_0 satisfies the optimality conditions for the model problem

$$\min\{c'x + \frac{1}{2}x'Cx \mid Ax \leq b\}, \tag{5.12}$$

if and only if

$$Ax_0 \leq b, \tag{5.13}$$

$$-\nabla f(x_0) = A'u, \ u \geq 0, \quad \text{and} \tag{5.14}$$

$$u'(Ax_0 - b) = 0. \tag{5.15}$$

Because $A' = [a_1, a_2, \ldots, a_m]$, (5.13) and (5.14) are identical to the primal and dual feasibility parts of Definition 5.3. Expanding condition (5.15) gives

$$u'(Ax_0 - b) = \sum_{i=1}^{m} u_i(a_i'x_0 - b_i).$$

From (5.14), each u_i is nonnegative. Furthermore, from (5.13) each $a_i'x_0 - b_i \leq 0$. This implies that $u'(Ax_0 - b)$ is a sum of terms each of which is nonpositive. Since the sum must be zero, it follows that each individual term must also be zero. This $u_i(a_i'x_0 - b_i) = 0$ for $i = 1, 2, \ldots, m$, in agreement with the complementary slackness part of Definition 5.3.

　　The major goal of this section is to show that the optimality conditions are necessary and sufficient for optimality. This is formulated in the following theorem.

Theorem 5.1 x_0 *is an optimal solution for the QP (5.9) if and only if x_0 satisfies the optimality conditions for QP (5.9).*

Proof: We will prove only sufficiency. The proof of necessity is a bit more complicated. See Mangasarian [18].

　　We use the optimality conditions for (5.9) as formulated in (5.13)–(5.15). Let x be any point in R. From Taylor's Theorem (Theorem 1.1),

$$f(x) = f(x_0) + \nabla f(x_0)'(x - x_0) + \frac{1}{2}(x - x_0)'C(x - x_0).$$

Since C is positive semidefinite, $(x - x_0)'C(x - x_0) \geq 0$ and thus

$$
\begin{aligned}
f(x) - f(x_0) &\geq \nabla f(x_0)'(x - x_0) \\
&\geq -u'A(x - x_0) && \text{(dual feasibility)} \\
&= u'Ax_0 - u'Ax \\
&= u'b - u'Ax && \text{(complementary slackness)} \\
&= u'(b - Ax).
\end{aligned}
$$

From dual feasibility, each component of u is nonnegative and because $x \in R$, each component of $b - Ax$ is also nonnegative. Thus $u'(b - Ax)$ is the inner product of two nonnegative vectors and must be nonnegative itself. Thus $f(x) - f(x_0) \geq 0$, or, $f(x_0) \leq f(x)$. Since x is an arbitrary point in R, we have shown $f(x_0) \leq f(x)$ for all $x \in R$ and by Definition 5.2, x_0 is indeed optimal for (5.9). □

In Section 1.2 we have seen optimality conditions for problems with linear equality constraints. In the present section we have formulated optimality conditions for problems with linear inequality constraints. We next formulate optimality conditions for a problem which has both linear inequality and equality constraints. In particular, we consider the problem

$$\min\{c'x + \frac{1}{2}x'Cx \mid A_1x \leq b_1, A_2x = b_2\}. \tag{5.16}$$

We proceed by rewriting (5.16) in the model form of (5.8), then applying the optimality conditions (5.13)–(5.15) and finally, simplifying the result. Note that $A_2x = b_2$ is equivalent to $A_2x \leq b_2$ and $-A_2x \leq -b_2$ so that (5.16) may be rewritten as

$$\begin{aligned}
\text{minimize}: \quad & c'x + \frac{1}{2}x'Cx \\
\text{subject to}: \quad & \begin{bmatrix} A_1 \\ A_2 \\ -A_2 \end{bmatrix} x \leq \begin{bmatrix} b_1 \\ b_2 \\ -b_2 \end{bmatrix}.
\end{aligned} \tag{5.17}$$

The QP (5.17) is now of the same algebraic form as (5.12) and we may use the optimality conditions (5.13)–(5.15) for it. These are

$$A_1x_0 \leq b_1, \quad A_2x_0 = b_2, \tag{5.18}$$

$$-\nabla f(x_0) = A_1'u_1 + A_2'u_2 - A_2'u_3, \quad u_1, u_2, u_3 \geq 0, \tag{5.19}$$

$$u_1'(A_1x_0 - b_1) = 0, \quad u_2'(A_2x_0 - b_2) = 0 \quad u_3'(-A_2x_0 + b_2) = 0 \tag{5.20}$$

Because in (5.18) $A_2x_0 = b_2$, the complementary slackness conditions (5.20) reduce to $u_1'(A_1x_0 - b_1) = 0$. Also, the dual feasibility condition (5.19) can be simplified:

$$-\nabla f(x_0) = A_1'u_1 + A_2'(u_2 - u_3), \quad u_1, u_2, u_3 \geq 0.$$

Although both u_2 and u_3 must be nonnegative, their difference can have any sign. Replacing u_1 with u and $u_2 - u_3$ with v, we have the optimality conditions for (5.16) are

$$\left.\begin{array}{c} A_1 x_0 \leq b_1, \quad A_2 x_0 = b_2, \\ -\nabla f(x_0) = A_1' u + A_2' v, \quad u \geq 0, \\ u'(A_1 x_0 - b_1) = 0 \end{array}\right\}. \qquad (5.21)$$

The key result here is that the multipliers associated with equality constraints are unconstrained in sign and there is no complementary slackness condition associated with equality constraints.

Summarizing our results, we have the following version of Theorem 5.1:

Theorem 5.2 x_0 *is an optimal solution for*

$$\min\{c'x + \frac{1}{2}x'Cx \mid A_1 x \leq b_1, A_2 x = b_2\}$$

if and only if

1. $A_1 x_0 \leq b_1, \quad A_2 x_0 = b_2,$
2. $-\nabla f(x_0) = A_1' u + A_2' v, \quad u \geq 0,$
3. $u'(A_1 x_0 - b_1) = 0.$

The following result gives conditions under which an optimal solution for a QP is uniquely determined and when the multipliers are also uniquely determined.

Theorem 5.3 *Let* x_0 *be an optimal solution for* $\min\{c'x + \frac{1}{2}x'Cx \mid Ax \leq b\}$ *and let* C *be positive definite. Then* x_0 *is the unique optimal solution. In addition, if the gradients of those constraints which are active at* x_0 *are linearly independent, then the multipliers for these constraints are uniquely determined.*

Proof: Suppose x_0 and x_1 are two distinct optimal solutions. Then $f(x_0) \leq f(x)$ and $f(x_1) \leq f(x)$ for all feasible x. Consequently,

$$f(x_0) = f(x_1). \qquad (5.22)$$

Since x_0 is an optimal solution, it must satisfy the optimality conditions; i.e., there exists an $m-$vector u_0 with

$$-\nabla f(x_0) = A'u_0, \quad u_0 \geq 0, \quad \text{and} \quad u_0'(Ax_0 - b) = 0. \quad (5.23)$$

From (5.23), it follows that

$$\begin{aligned}
\nabla f(x_0)'(x_1 - x_0) &= -u_0'Ax_1 + u_0'b \\
&= u_0'(b - Ax_1).
\end{aligned} \quad (5.24)$$

From Taylor's Theorem,

$$f(x_1) = f(x_0) + \nabla f(x_0)'(x_1 - x_0) + \frac{1}{2}(x_1 - x_0)'C(x_1 - x_0). \quad (5.25)$$

Substitution of (5.22) and (5.24) into (5.25) gives

$$u_0'(b - Ax_1) + \frac{1}{2}(x_1 - x_0)'C(x_1 - x_0) = 0.$$

Because x_1 is feasible, $b - Ax_1 \geq 0$. Furthermore, from (5.23) $u_0 \geq 0$. Thus both terms in the above sum are nonnegative. Since they sum to zero, they must both be zero. In particular, we must have $(x_1 - x_0)'C(x_1 - x_0) = 0$. But because C is positive definite, this implies $x_1 = x_0$. The assumption that there are two distinct optimal solutions leads to a contradiction and is therefore false. Therefore, x_0 is the unique optimal solution.

To show uniqueness of the multipliers, assume there are two multiplier vectors u_0 and u_1. Letting $I(x_0) = \{i \mid a_i'x_0 = b_i\}$ denote the set of indices of constraints active at x_0, we have

$$-\nabla f(x_0) = \sum_{i \in I(x_0)} (u_0)_i a_i = \sum_{i \in I(x_0)} (u_1)_i a_i$$

and so

$$\sum_{i \in I(x_0)} [(u_0)_i - (u_1)_i] a_i = 0.$$

Because the gradients of those constraints active at x_0 are linearly independent,

$$(u_0)_i = (u_1)_i, \quad \text{for all } i \in I(x_0).$$

Since for all i not in $I(x_0)$, $(u_0)_i = (u_1)_i = 0$, we have $u_0 = u_1$, and u_0 is indeed uniquely determined. $\square$

5.5   Exercises

5.1 Let $f(x) = -196x_1 - 103x_2 + \frac{1}{2}(37x_1^2 + 13x_2^2 + 32x_1x_2)$.

(a) Use the eigenvectors and eigenvalues of the Hessian matrix to sketch the level sets of f.

(b) Give a graphical solution to the problem of minimizing f subject to:

$$
\begin{array}{rcrcll}
- & x_1 & + & x_2 & \leq & 0, \quad (1) \\
  & x_1 &   &   & \leq & 3, \quad (2) \\
  &   & - & x_2 & \leq & 0. \quad (3)
\end{array}
$$

Show graphically that the optimality conditions are satisfied.

5.2 As the Director of Optimization for Waterloo International Investments, you are evaluating various QP codes offered by a variety of vendors. You decide to test the QP codes on problems of the form $\min\{c'x + \frac{1}{2}x'Cx \mid Ax \leq b\}$, where

$$
C = \begin{bmatrix} 1 & 0 & 0 \\ 0 & 2 & 1 \\ 0 & 1 & 3 \end{bmatrix}, \quad A = \begin{bmatrix} 1 & 1 & 1 \\ 2 & 0 & 3 \end{bmatrix}, \quad b = \begin{bmatrix} 3 \\ 5 \end{bmatrix},
$$

and various choices for c will determine the specific test problems. When you use $c = (-4, -4, -8)'$, the QP code gives an optimal solution of $x^* = (1, 1, 1)'$. When you use $c = (0, -4, -2)'$, the QP code also gives an optimal solution of $x^* = (1, 1, 1)'$ and when $c = (1, 1, 1)'$, the QP code again claims that $x^* = (1, 1, 1)'$ is optimal. Give a detailed analysis of the results.

5.3 For the problem $\min\{c'x + \frac{1}{2}x'Cx \mid Ax \leq b\}$, let $x^* = (1, 1, 1)'$,

$$
C = \begin{bmatrix} 1 & 0 & 0 \\ 0 & 3 & 1 \\ 0 & 1 & 2 \end{bmatrix}, \quad A = \begin{bmatrix} 1 & 1 & 2 \\ 3 & 0 & 3 \end{bmatrix}, \quad b = \begin{bmatrix} 4 \\ 6 \end{bmatrix}.
$$

For each of the following values of c, either prove that x^* is optimal or prove that it is not optimal: (a) $c = (-5, -5, -6)'$, (b) $c = (-3, -3, -4)'$, (c) $c = (-5, -5, -5)'$.

5.4 Use the eigenvector-eigenvalue method to sketch level sets for $f(x) = c'x + \frac{1}{2}x'Cx$ where

$$c = \begin{bmatrix} -4 \\ -4 \end{bmatrix}, \quad C = \begin{bmatrix} 3 & -1 \\ -1 & 3 \end{bmatrix}.$$

5.5 Consider the problem

$$\text{minimize}: \quad -67x_1 - 59x_2 + \frac{13}{2}x_1^2 + 5x_1x_2 + \frac{13}{2}x_2^2$$

$$\begin{array}{rcll} \text{subject to}: \quad x_1 + x_2 & \leq & 4, & (1) \\ -x_1 & \leq & 0, & (2) \\ -x_2 & \leq & 0. & (3) \end{array}$$

Solve this problem either by enumerating possible active sets or by graphing it to determine the active constraints and then use the optimality conditions to locate the precise optimal solution.

5.6 Use the optimality conditions for the problem $\min\{-\mu'x \mid x \geq 0, \; l'x = 1\}$ to verify that the optimal solution determined in Example 1.4 is indeed optimal.

5.7 Consider the following portfolio optimization problem:

$$\min\{-t\mu'x + \frac{1}{2}x'\Sigma x \mid l'x = 1, \; x \geq 0\}.$$

Use the optimality conditions to show that for all t sufficiently large (state how large) the efficient portfolios have all holdings in just one asset (state which one) and zero holdings in the remaining $n-1$ assets. Assume that $\mu_1 < \mu_2 \cdots < \mu_n$.

5.8 Give an example of a QP for which the optimal solution is not unique.

5.9 Give an example of a QP for which the multipliers are not unique.

5.10 Consider the following portfolio optimization problem:

$$\min\{-t\mu'x + \frac{1}{2}x'\Sigma x \mid l'x = 1, \; x \geq 0\},$$

where $\Sigma = \text{diag}(\sigma_1, \sigma_2, \ldots, \sigma_n)$ and $\mu_1 < \mu_2 \cdots < \mu_n$. Use the optimality conditions for it to answer the following questions.

(a) Determine the efficient portfolios for it as explicit functions of t for all nonnegative t sufficiently small.

(b) What is the largest value of t (call it t_1) such that this remains valid?

(c) What happens when t is increased a small amount beyond t_1? Prove it.

5.11 Consider the problem

$$\min\{\, f(x) \mid Ax \leq b \,\},$$

where $f(x)$ is a nonlinear function of the n−vector x. The optimality conditions for this problem are

(a) $Ax \leq b$,

(b) There exists an m−vector u with $-\nabla f(x) = A'u$, $u \geq 0$,

(c) $u'(Ax - b) = 0$.

Assume the Hessian matrix, $H(x)$ is positive definite for all x. Show that if x is a point for which the optimality conditions are satisfied, then x is indeed an optimal solution for the given problem. (*Hint:* Try basing your proof on the proof of Theorem 5.1.)

Chapter 6

A QP Solution Algorithm

In previous chapters, we have used a model portfolio optimization problem with only a budget constraint. Practical portfolio problems may have nonnegativity constraints ($x \geq 0$) on the asset holdings to preclude the possibility of short sales. These are lower bound constraints. Portfolio managers may also include upper bound constraints so that any change in asset holdings is not too large. Practitioners may also impose sector constraints which require that the portfolio's holdings in particular industry sectors, for example oil related sectors, does not exceed a certain percentage. Transaction costs may also be included by introducing new variables which are restricted to be nonnegative. Thus a realistic portfolio optimization problem must include general linear inequality constraints.

A QP solution algorithm is a method used to minimize a quadratic function of many variables subject to linear inequality and equality constraints. There are many such methods and in Best [2], it was shown that several of these methods are equivalent in that if they are initialized with the identical starting point and ties were resolved using a common rule, then the same sequence of points would be generated by all such methods. The methods differ only in the way they solve certain linear equations at each iteration. The equivalency results were obtained by first formulating a generic QP algorithm (called QP Algo-

107

rithm A) which left unspecified the method by which the linear equations were solved. Then each other method was shown to be equivalent to QP Algorithm A. The formulation of QP Algorithm A is quite simple, as it just states that certain linear equations are to be solved but does not go into the details as to how it is to be done. We shall develop QP Algorithm A in this chapter. By assuming the Hessian matrix of the quadratic objective function is positive definite, the development of this method will be made even simpler.

6.1   QPSolver: A QP Solution Algorithm

In this section we will formulate a QP solution algorithm for the model problem

$$\min\{c'x + \frac{1}{2}x'Cx \mid a_i'x \leq b_i, \ i = 1, 2, \ldots, m\}. \tag{6.1}$$

Our model problem thus has m linear inequality constraints. Let $R = \{x \mid a_i'x \leq b_i, i = 1, 2, \ldots, m\}$ denote the feasible region for (6.1). We will assume throughout this section that C is symmetric and positive definite. Consequently, the optimality conditions of the previous section are both necessary and sufficient for optimality. The assumption that C is positive definite is unduly restrictive and can be weakened to C being positive semidefinite. We use the positive definite assumption here to present the key ideas of a quadratic programming algorithm.

As with linear programming, the key idea for quadratic programming is to formulate an algorithm which will terminate in a *finite* number of steps. Recall from linear programming that x_0 is an extreme point for R if $x_0 \in R$ and there are n constraints having linearly independent gradients active at x_0. The number of extreme points is finite and linear programming algorithms (notably the simplex method) move from one extreme point to an adjacent extreme point in such a way that the objective function decreases and thus solves the problem in a finite number of steps. We want to generalize this idea to quadratic programming problems and as such, we need a generalization of an extreme point. To this end, for any $x_0 \in R$ let

$$I(x_0) = \{\, i \mid a_i' x_0 = b_i \,\}.$$

$I(x_0)$ is simply the set of indices of those constraints which are active at x_0. The generalization of an extreme point is formulated in the following definition.

Definition 6.1 *The point x_0 is a* **quasi-stationary point** *for (6.1) if (1) $x_0 \in R$ and (2) x_0 is optimal for*

$$\min\{c'x + \frac{1}{2}x'Cx \mid a_i'x = b_i, \text{ all } i \in I(x_0)\}.$$

Definition 6.2 *The point x_0 is a* **nondegenerate quasi-stationary point** *for (6.1) if it is a quasi-stationary point for (6.1) and the gradients of those constraints active at x_0 are linearly independent. The point x_0 is a* **degenerate quasi-stationary point** *for (6.1) if it is a quasi-stationary point for (6.1) and the gradients of those constraints active at x_0 are linearly dependent.*

If x_0 is an extreme point for (6.1) then by definition $x_0 \in R$, and furthermore the feasible region for the problem in part (2) of the definition of a quasi-stationary point is the single point x_0 and therefore x_0 is optimal for that problem. Consequently, an extreme point is a special case of a quasi-stationary point, or alternatively, the notion of a quasi-stationary point generalizes that of an extreme point. In addition, there are only finitely many quasi-stationary points. There can be no more than 2^m (the number of subsets of $\{1, 2, \ldots, m\}$) quasi-stationary points. Finally, it follows directly from the optimality conditions (Theorem 5.1) that an optimal solution for (6.1) is a quasi-stationary point. The basic idea of a quadratic programming solution algorithm should now be quite clear. We move from one quasi-stationary point to another, in such a way as to decrease the objective function. There are only finitely many quasi-stationary points and one of them is the optimal solution. Consequently, the optimal solution will be obtained in finitely many steps. Analogous to the nondegeneracy assumption usually made for linear programming, we will make the assumption that each quasi-stationary point is nondegenerate.

Quasi-stationary points are illustrated in the following example.

Example 6.1

$$\begin{array}{rlrl}
\text{minimize:} & \quad -10x_1 - 4x_2 + x_1^2 + x_2^2 & & \\
\text{subject to:} & \quad x_1 & \leq \;\; 4, & \quad (1) \\
& \quad -x_2 \leq -1, & & \quad (2) \\
& \quad -x_1 & \leq -1, & \quad (3) \\
& \quad x_2 \leq \;\; 3. & & \quad (4)
\end{array}$$

The geometry of Example 6.1 is illustrated in Figure 6.1. Note that we have used y_1 and y_2 to denote the coordinate axes and $x_0, x_1, \ldots, x_8$ to denote potential quasi-stationary points (vectors). Level sets for the objective function are circles centered at $(5, 2)'$. Because x_1, x_2, x_3 and x_4 are extreme points, they are also quasi-stationary points. Let $f(x)$ denote the objective function for this example. Minimizing $f(x)$ subject to constraint 3 holding as an equality constraint gives x_5. Since $x_5 \in R$, x_5 is indeed a quasi-stationary point. Similarly, x_6 is also a quasi-stationary point (and indeed, x_6 is also the optimal solution.) Minimizing $f(x)$ subject to constraint 4 holding as an equality constraint gives x_7. Since $x_7 \notin R$, x_7 is not a quasi-stationary point. Similarly, x_8 is not a quasi-stationary point.   ◊

We proceed to solve (6.1) by solving a sequence of equality constrained problems of the form

$$\min\{c'x + \frac{1}{2}x'Cx \mid Ax = b\}, \tag{6.2}$$

where the equality constraints $Ax = b$ are a subset of those of (6.1) which are temporarily required to hold as equalities. Although we could solve (6.2) directly, it is convenient to solve it in a slightly different form. Let $f(x) = c'x + \frac{1}{2}x'Cx$. Let x_0 be a feasible point for (6.2), let $g_0 = \nabla f(x_0)$ and we wish to write the optimal solution for (6.2) in the form $x_0 - s_0$ so that $-s_0$ has the interpretation of being a search direction emanating from x_0 and pointing to the optimal solution for (6.2). Now x_0 is known and we wish to obtain s_0. By Taylor's Theorem,

$$f(x_0 - s_0) = f(x_0) - g_0's_0 + \frac{1}{2}s_0'Cs_0.$$

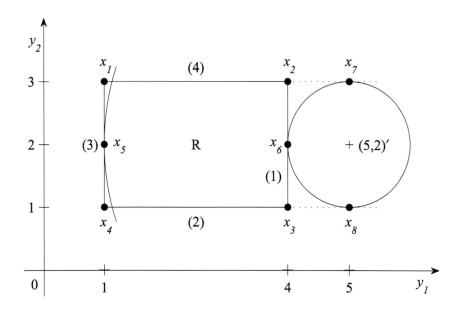

Figure 6.1   Quasi-stationary Points for Example 6.1.

Since $Ax_0 = b$ and we want $A(x_0 - s_0) = b$, we must require $As_0 = 0$. Thus solving (6.2) in terms of s_0 is equivalent to solving

$$\min\{-g_0's_0 + \frac{1}{2}s_0'Cs_0 \mid As_0 = 0\}. \tag{6.3}$$

From Theorem 1.2, necessary and sufficient conditions for optimality for (6.3) are:

$$g_0 - Cs_0 = A'v, \quad As_0 = 0,$$

or, in partitioned matrix form,

$$\begin{bmatrix} C & A' \\ A & 0 \end{bmatrix} \begin{bmatrix} s_0 \\ v \end{bmatrix} = \begin{bmatrix} g_0 \\ 0 \end{bmatrix}.$$

Summarizing, we have

Lemma 6.1 *Let x_0 satisfy $Ax_0 = b$. Then $x_0 - s_0$ is optimal for $\min\{c'x + \frac{1}{2}x'Cx \mid Ax = b\}$ if and only if s_0 satisfies the linear equations*

$$\begin{bmatrix} C & A' \\ A & 0 \end{bmatrix} \begin{bmatrix} s_0 \\ v_0 \end{bmatrix} = \begin{bmatrix} g_0 \\ 0 \end{bmatrix},$$

where $g_0 = \nabla f(x_0)$.

Let

$$H = \begin{bmatrix} C & A' \\ A & 0 \end{bmatrix}. \tag{6.4}$$

Matrices of the form of H will play a critical role in the development of a quadratic programming algorithm. Nonsingularity of such matrices when A has full row rank is established in the following lemma.

Lemma 6.2 *Suppose A has full row rank. Then H defined by (6.4) is nonsingular.*

Proof: Suppose to the contrary that H is singular. Then there exist vectors s and v, not both zero such that

$$\begin{bmatrix} C & A' \\ A & 0 \end{bmatrix} \begin{bmatrix} s \\ v \end{bmatrix} = \begin{bmatrix} 0 \\ 0 \end{bmatrix}.$$

Writing out the two partitions gives

$$Cs + A'v = 0, \tag{6.5}$$

$$As = 0. \tag{6.6}$$

Taking the inner product of both sides of (6.5) with s gives

$$s'Cs + s'A'v = 0.$$

But from (6.6), $s'A' = 0$, so that $s'Cs = 0$ and since C is positive definite this implies that $s = 0$. Thus (6.5) reduces to $A'v = 0$. But

A has full row rank, so $v = 0$. Thus both s and v are zero. This contradiction establishes that H is indeed nonsingular.   □

We can now begin to formulate a solution method for (6.1). Sometimes it is convenient to write (6.1) more compactly as

$$\min\{c'x + \frac{1}{2}x'Cx \mid Ax \leq b\}. \tag{6.7}$$

Let x_0 be a known feasible point for (6.1). Let A_0' denote the submatrix of A corresponding to those constraints active at x_0 and let b_0 denote the corresponding subvector of b. So for example, if constraints $5, 10$ and 15 were active at x_0 we would have

$$A_0 = \begin{bmatrix} a_5' \\ a_{10}' \\ a_{15}' \end{bmatrix} \quad \text{and} \quad b_0 = \begin{bmatrix} b_5 \\ b_{10} \\ b_{15} \end{bmatrix}. \tag{6.8}$$

Assume that A_0 has full row rank and define

$$H_0 = \begin{bmatrix} C & A_0' \\ A_0 & 0 \end{bmatrix}.$$

According to Lemma 6.2, H_0 is nonsingular and we can solve the linear equations

$$H_0 \begin{bmatrix} s_0 \\ v_0 \end{bmatrix} = \begin{bmatrix} g_0 \\ 0 \end{bmatrix}, \tag{6.9}$$

to obtain the search direction s_0.

Let σ be a nonnegative scalar and consider the function $f(x_0 - \sigma s_0)$. Then $f(x_0 - \sigma s_0)$ is decreasing function of σ for $0 \leq \sigma \leq 1$ and attains its minimum for $\sigma = 1$. (See Exercise 6.2.) However, in the determination of s_0, those constraints which are inactive at x_0 were ignored so that the point $x_0 - s_0$ may not be feasible for (6.1). We would like to know the largest value of σ such that $x_0 - \sigma s_0$ remains feasible. For a typical constraint i, we require

$$a_i'(x_0 - \sigma s_0) \leq b_i,$$

or,

$$-\sigma a_i' s_0 \leq b_i - a_i' x_0. \tag{6.10}$$

If $a_i' s_0 \geq 0$ then (6.10) imposes no restriction because increasing σ results in the left-hand side of (6.10) decreasing. If $a_i' s_0 < 0$, then σ must satisfy

$$\sigma \leq \frac{a_i' x_0 - b_i}{a_i' s_0}, \tag{6.11}$$

where we have divided (6.10) by the positive quantity $-a_i' s_0$. Now (6.11) must be satisfied for all i with $a_i' s_0 < 0$. The largest such value for σ is called the *maximum feasible step size* $\hat{\sigma}_0$ and is given by the formula

$$
\begin{aligned}
\hat{\sigma}_0 &= \min\left\{ \frac{a_i' x_0 - b_i}{a_i' s_0} \;\middle|\; \text{all } i = 1, 2, \ldots, m \text{ with } a_i' s_0 < 0 \right\} \\
&= \frac{a_l' x_0 - b_l}{a_l' s_0}, \tag{6.12}
\end{aligned}
$$

where l is the index for which the minimum is attained.

As an example, consider the constraints of Example 6.1. Take $x_0 = (1, 2)'$ and $s_0 = (-1, -1)'$. The maximum feasible step size for s_0 is

$$\hat{\sigma}_0 = \min\left\{ \frac{-3}{-1}, -, -, \frac{-1}{-1} \right\} = 1,$$

and $l = 4$. The two dashes indicate that $a_2' s_0 \geq 0$ and $a_3' s_0 \geq 0$. Note that at $x_0 - \hat{\sigma}_0 s_0$, constraint $l = 4$ becomes active and that is the constraint for which the minimum occurred in the computation of $\hat{\sigma}_0$.

We now know that $x_0 - \sigma s_0 \in R$ for all σ such that $0 \leq \sigma \leq \hat{\sigma}_0$ and that $f(x_0 - \sigma s_0)$ is decreasing for $0 \leq \sigma \leq 1$. The greatest decrease in f compatible with maintaining feasibility is obtained by setting

$$\sigma_0 = \min\{\hat{\sigma}_0, 1\}, \tag{6.13}$$

and our next estimate of an optimal solution for (6.1) is

$$x_1 = x_0 - \sigma_0 s_0. \tag{6.14}$$

The analysis now continues according to two cases: $\sigma_0 < 1$ (Case 1) and $\sigma_0 = 1$ (Case 2).

Case 1: $\sigma_0 < 1$

In this case, some previously inactive constraint has become active at x_1. According to (6.12), the index of this constraint is l. We can now proceed by adding constraint l to the active set and repeating. That is, compute $g_1 = \nabla f(x_1)$, form A_1 from A_0 by adding a'_l as the last row, forming

$$H_1 = \begin{bmatrix} C & A'_1 \\ A_1 & 0 \end{bmatrix},$$

and then solving

$$H_1 \begin{bmatrix} s_1 \\ v_1 \end{bmatrix} = \begin{bmatrix} g_1 \\ 0 \end{bmatrix}$$

to obtain s_1, compute the maximum feasible step size $\hat{\sigma}_1$ for x_1 and s_1, compute $\sigma_1 = \min\{\hat{\sigma}_1, 1\}$ and set $x_2 = x_1 - \sigma_1 s_1$.

Note that a_l cannot be linearly dependent on the columns of A'_0 because if it were, there would be a nonzero vector w satisfying $a_l = A'_0 w$. Taking the inner product of both sides of this last equation with s_0 and using the fact that $A_0 s_0 = 0$ gives $a'_l s_0 = 0$. But by definition of the maximum feasible step size (6.12), $a'_l s_0 < 0$. This contradiction establishes that A_1 does indeed have full row rank.

If $\sigma_1 < 1$, we could again form A_2 by adding the gradient of the new active constraint and proceeding as before. In general for $j = 0, 1, \ldots$ we would start with x_j, A_j, compute $g_j = \nabla f(x_j)$, form

$$H_j = \begin{bmatrix} C & A'_j \\ A_j & 0 \end{bmatrix},$$

then solve

$$H_j \begin{bmatrix} s_j \\ v_j \end{bmatrix} = \begin{bmatrix} g_j \\ 0 \end{bmatrix}$$

for s_j. Next we would compute the maximum feasible step size $\hat{\sigma}_j$ for x_j and s_j, set $\sigma_j = \min\{\hat{\sigma}_j, 1\}$ and set $x_{j+1} = x_j - \sigma_j s_j$. If $\sigma_j < 1$, we would form A_{j+1} from A_j by augmenting it with the gradient (transposed) of the new active constraint and once again repeat. Let

p denote the number of constraints active at x_0. Each successive time the condition $\sigma_j < 1$ occurs, the number of active constraints increases by one. Their gradients are linearly independent so no more than n constraints can be active. Thus, Case 2 must occur in at most $n - p$ iterations.

Case 2: $\sigma_0 = 1$
As discussed in Case 1, this case could occur after several occurrences of Case 1, but for ease of notation we will assume it occurs right at iteration 0. For Case 2, it is easy to see that by construction x_1 is a quasi-stationary point. $x_1 = x_0 - s_0$ was constructed to be optimal for

$$\min\{c'x + \frac{1}{2}x'Cx \mid A_0 x = \hat{b}_0\},$$

and because $\hat{\sigma}_0 \geq 1$, $x_1 \in R$. Thus x_1 is indeed a quasi-stationary point.

We next show that x_1 satisfies all of the optimality conditions for (6.1) with the possible exception of the nonnegativity of the multipliers. Because $x_1 = x_0 - s_0$,

$$\begin{aligned} g_1 &= c + Cx_1 \\ &= c + C(x_0 - s_0) \\ &= c + Cx_0 - Cs_0 \\ &= g_0 - Cs_0 \end{aligned}$$

so that

$$g_0 = g_1 + Cs_0. \tag{6.15}$$

From (6.9),
$$Cs_0 + A_0'v_0 = g_0.$$

Using (6.15) in this last gives

$$g_1 = A_0'v_0. \tag{6.16}$$

This shows that $\nabla f(x_1)$ is a linear combination of the gradients of those constraints active at x_1.

For the situation in (6.8), we would have

$$-\nabla f(x_1) = (-(v_0)_1)a_5 + (-(v_0)_2)a_{10} + (-(v_0)_3)a_{15},$$

If this problem had $m = 20$ constraints, we could define the multiplier vector u by

$$u_i = 0, \quad i = 1, 2, \ldots, 20, \quad i \neq 5, 10, 15,$$

$$u_5 = -(v_0)_1, \quad u_{10} = -(v_0)_2, \quad u_{15} = -(v_0)_3.$$

Thus, the point x_1 together with the multiplier vector u, satisfies all of the optimality conditions with the possible exception of the nonnegativity requirement for u.

Returning to the general situation (6.16), we can construct a multiplier vector u from $-v_0$ in a similar manner. If $u \geq 0$, then x_1 together with u satisfies all of the optimality conditions for (6.1) and x_1 is an optimal solution. One way to check this is to compute

$$u_k = \min\{u_i : i = 1, 2, \ldots, m\}. \tag{6.17}$$

If $u_k \geq 0$, then $u \geq 0$ and x_1 is optimal. Otherwise $u_k < 0$ and we continue by deleting constraint k from the active set; i.e., we obtain A_1 from A_0 by deleting that row of A_0 which contained a_k'. We now continue to locate the next quasi-stationary point with objective function value strictly less than that for x_1. By deleting constraint k from the active set, we need to be assured that constraint k does indeed become inactive at points of the form $x_1 - \sigma s_1$ for $\sigma \geq 0$. To see why this will be the case, observe that from (6.16),

$$g_1 = A_0' v_0, \tag{6.18}$$

by construction of s_1,

$$A_1 s_1 = 0 \tag{6.19}$$

and since A_1 is obtained from A_0 by dropping constraint k,

$$A_0' = [A_1', a_k]. \tag{6.20}$$

Equations (6.18) and (6.20) imply

$$g_1 = [A_1', a_k]v_0 \qquad (6.21)$$

Taking the inner product of both sides of (6.21) with s_1 and using 6.19 gives

$$g_1's_1 = v_0' \begin{bmatrix} A_1s_1 \\ a_k's_1 \end{bmatrix} = v_0' \begin{bmatrix} 0 \\ a_k's_1 \end{bmatrix} = -u_k(a_k's_1), \qquad (6.22)$$

where the last equality in (6.22) is a consequence of the full dimensional vector of multipliers, u being obtained from $-v_0$. Summarizing, we have

$$-g_1's_1 = (a_k's_1)u_k. \qquad (6.23)$$

But $g_1's_1 > 0$ (Exercise 6.1(b)) and k was chosen such that $u_k < 0$. Consequently,

$$a_k's_1 > 0. \qquad (6.24)$$

But then

$$\begin{aligned} a_k'(x_1 - \sigma s_1) &= a_k'x_1 - \sigma a_k's_1 \\ &= b_k - \sigma a_k's_1 \end{aligned}$$

and constraint k is inactive for $\sigma > 0$.

If we modify our model problem (6.1) to include linear equality constraints, these can be handled in a very straightforward way. All equality constraints must be in the active set at the beginning of the algorithm and they are never allowed to become inactive.

In giving a detailed formulation of the algorithm, it is useful to employ an index set K_j containing the indices of the active constraints at iteration j. Note that if the problem contains linear equality constraints, all of their indices must be in K_0.

In order to initiate the algorithm, we require a feasible point x_0 and $K_0 \subseteq I(x_0)$ to satisfy the following assumption.

Assumption 6.1.

a_i, all $i \in K_0$, are linearly independent.

As a consequence of Assumption 6.1, it follows from Lemma 6.2 that H_0 is nonsingular.

A detailed statement of the algorithm follows.

QPSOLVER

Model Problem:

$$
\begin{aligned}
\text{minimize:} \quad & c'x + \tfrac{1}{2}x'Cx \\
\text{subject to:} \quad & a_i'x \leq b_i, \quad i = 1, 2, \ldots, m \\
& a_i'x = b_i, \quad i = m+1, m+2, \ldots, m+q.
\end{aligned}
$$

Initialization:

Start with any feasible point x_0 and active set $K_0 \subseteq I(x_0)$ such that Assumption 6.1 is satisfied and each of $m+1, m+2, \ldots, m+q$ is in K_0. Compute $f(x_0) = c'x_0 + \tfrac{1}{2}x_0'Cx_0$, $g_0 = c + Cx_0$, and set $j = 0$.

Step 1:   **Computation of Search Direction s_j.**

Let

$$
H_j = \begin{bmatrix} C & A_j' \\ A_j & 0 \end{bmatrix}.
$$

Compute the search direction s_j and multipliers v_j as the solution of the linear equations

$$
H_j \begin{bmatrix} s_j \\ v_j \end{bmatrix} = \begin{bmatrix} g_j \\ 0 \end{bmatrix},
$$

and go to Step 2.

Step 2:   Computation of Step Size σ_j.

If $a_i's_j \geq 0$ for $i = 1, 2, \ldots, m$, set $\hat{\sigma}_j = +\infty$. Otherwise, compute the smallest index l and $\hat{\sigma}_j$ such that

$$\hat{\sigma}_j = \frac{a_l'x_j - b_l}{a_l's_j} = \min \left\{ \frac{a_i'x_j - b_i}{a_i's_j} \mid \text{all } i \notin K_j \text{ with } a_i's_j < 0 \right\}.$$

Set $\sigma_j = \min\{\hat{\sigma}_j, 1\}$ and go to Step 3.

Step 3:   Update.

Set $x_{j+1} = x_j - \sigma_j s_j$, $g_{j+1} = c + C x_{j+1}$, and $f(x_{j+1}) = c'x_{j+1} + \frac{1}{2}x_{j+1}'C x_{j+1}$. If $\sigma_j < 1$, go to Step 3.1. Otherwise, go to Step 3.2.

Step 3.1:

Set $K_{j+1} = K_j + \{l\}$, form A_{j+1} and H_{j+1}, replace j with $j + 1$ and go to Step 1.

Step 3.2:

Compute the multiplier vector u_{j+1} from $-v_j$ and K_j and compute k such that

$$(u_{j+1})_k = \min\{(u_{j+1})_i \mid i = 1, 2, \ldots, m\}.$$

If $(u_{j+1})_k \geq 0$, then stop with optimal solution x_{j+1}. Otherwise, set $K_{j+1} = K_j - \{k\}$, form A_{j+1} and H_{j+1}, replace j with $j + 1$ and go to Step 1.

Note that in Step 3.2 of QPSolver only the multipliers for active inequality constraints are used. The multipliers for equality constraints are not used and so all equality constraints are active at the start of the algorithm and remain active for all iterations.

We illustrate QPSolver by applying it to the following example.

Example 6.2

$$\begin{aligned}
\text{minimize:} \quad & -10x_1 - 4x_2 + x_1^2 + x_2^2 \\
\text{subject to:} \quad & x_1 && \leq \;\; 4, && (1) \\
& && -x_2 \leq -1, && (2) \\
& -x_1 && \leq -1, && (3) \\
& && x_2 \leq \;\; 3. && (4)
\end{aligned}$$

Here

$$c = \begin{bmatrix} -10 \\ -4 \end{bmatrix} \quad \text{and} \quad C = \begin{bmatrix} 2 & 0 \\ 0 & 2 \end{bmatrix}.$$

Initialization:

$$x_0 = \begin{bmatrix} 1 \\ 3 \end{bmatrix}, \quad K_0 = \{\, 3,\, 4\,\}, \quad A_0 = \begin{bmatrix} a_3' \\ a_4' \end{bmatrix} = \begin{bmatrix} -1 & 0 \\ 0 & 1 \end{bmatrix},$$

$$H_0 = \begin{bmatrix} 2 & 0 & -1 & 0 \\ 0 & 2 & 0 & 1 \\ -1 & 0 & 0 & 0 \\ 0 & 1 & 0 & 0 \end{bmatrix}, \quad f(x_0) = -12, \quad g_0 = \begin{bmatrix} -8 \\ 2 \end{bmatrix}, \quad j = 0.$$

Iteration 0

Step 1: Solving $H_0 \begin{bmatrix} s_0 \\ v_0 \end{bmatrix} = \begin{bmatrix} g_0 \\ 0 \end{bmatrix}$ gives $s_0 = \begin{bmatrix} 0 \\ 0 \end{bmatrix}$ and

$$v_0 = \begin{bmatrix} 8 \\ 2 \end{bmatrix}.$$

Step 2: $\hat{\sigma}_0 = +\infty,\ \sigma_0 = \min\{1, +\infty\} = 1$.

Step 3: $x_1 = \begin{bmatrix} 1 \\ 3 \end{bmatrix}, \quad g_1 = \begin{bmatrix} -8 \\ 2 \end{bmatrix}, \quad f(x_1) = -12.$

Transfer to Step 3.2.

Step 3.2: $u_1 = (0,\ 0,\ -8,\ -2)'$,

$(u_1)_3 = \min\{\,0,\ 0,\ -8,\ -2\,\} = -8, \qquad k = 3,$

$$K_1 = \{\,4\,\}, \quad A_1 = \begin{bmatrix} 0 & 1 \end{bmatrix}, \quad H_1 = \begin{bmatrix} 2 & 0 & 0 \\ 0 & 2 & 1 \\ 0 & 1 & 0 \end{bmatrix},$$

$j = 1.$

Iteration 1

Step 1:   Solving $H_1 \begin{bmatrix} s_1 \\ v_1 \end{bmatrix} = \begin{bmatrix} g_1 \\ 0 \end{bmatrix}$ gives $s_1 = \begin{bmatrix} -4 \\ 0 \end{bmatrix}$ and

$v_1 = \begin{bmatrix} 2 \end{bmatrix}.$

Step 2: $\hat{\sigma}_1 = \min\left\{\dfrac{-3}{-4},\ -,\ -,\ -\right\} = \dfrac{3}{4}, \qquad l = 1,$

$\sigma_1 = \min\left\{\dfrac{3}{4},\ 1\right\} = \dfrac{3}{4}.$

Step 3: $x_2 = \begin{bmatrix} 1 \\ 3 \end{bmatrix} - \dfrac{3}{4}\begin{bmatrix} -4 \\ 0 \end{bmatrix} = \begin{bmatrix} 4 \\ 3 \end{bmatrix}, \quad g_2 = \begin{bmatrix} -2 \\ 2 \end{bmatrix},$

$f(x_2) = -27.$ Transfer to Step 3.1.

Step 3.1: $K_2 = \{\,4,\ 1\,\}, \qquad A_2 = \begin{bmatrix} 0 & 1 \\ 1 & 0 \end{bmatrix},$

$$H_2 = \begin{bmatrix} 2 & 0 & 0 & 1 \\ 0 & 2 & 1 & 0 \\ 0 & 1 & 0 & 0 \\ 1 & 0 & 0 & 0 \end{bmatrix}, \qquad j = 2.$$

Iteration 2

Step 1:   Solving $H_2 \begin{bmatrix} s_2 \\ v_2 \end{bmatrix} = \begin{bmatrix} g_2 \\ 0 \end{bmatrix}$ gives $s_2 = \begin{bmatrix} 0 \\ 0 \end{bmatrix}$ and

$$v_2 = \begin{bmatrix} 2 \\ -2 \end{bmatrix}.$$

Step 2: $\hat{\sigma}_2 = +\infty, \sigma_2 = \min\{+\infty, 1\} = 1.$

Step 3: $x_3 = \begin{bmatrix} 4 \\ 3 \end{bmatrix}, \quad g_3 = \begin{bmatrix} -2 \\ 2 \end{bmatrix}, \quad f(x_3) = -27.$

Transfer to Step 3.2.

Step 3.2: $u_3 = (2,\ 0,\ 0,\ -2)',$

$(u_3)_4 = \min\{2,\ 0,\ 0,\ -2\} = -2, \quad k = 4,$

$$K_3 = \{1\}, \quad A_3 = [\ 1 \quad 0\ ], \quad H_3 = \begin{bmatrix} 2 & 0 & 1 \\ 0 & 2 & 0 \\ 1 & 0 & 0 \end{bmatrix},$$

$j = 3.$

Iteration 3

Step 1:   Solving $H_3 \begin{bmatrix} s_3 \\ v_3 \end{bmatrix} = \begin{bmatrix} g_3 \\ 0 \end{bmatrix}$ gives $s_3 = \begin{bmatrix} 0 \\ 1 \end{bmatrix}$ and

$v_3 = [\ -2\ ].$

Step 2: $\hat{\sigma}_3 = \min \left\{ -, \dfrac{-2}{-1}, -, - \right\} = 2, \quad l = 2,$

$\sigma_3 = \min\{2,\ 1\} = 1.$

Step 3: $x_4 = \begin{bmatrix} 4 \\ 3 \end{bmatrix} - \begin{bmatrix} 0 \\ 1 \end{bmatrix} = \begin{bmatrix} 4 \\ 2 \end{bmatrix}, \quad g_4 = \begin{bmatrix} -2 \\ 0 \end{bmatrix},$

$f(x_4) = -28.$ Transfer to Step 3.2.

Step 3.2: $u_4 = (2,\ 0,\ 0,\ 0)',$

$(u_4)_2 = \min\{2,\ 0,\ 0,\ 0\} = 0, \quad k = 2.$

$(u_4)_2 \geq 0$; stop with optimal solution $x_4 = (4,\ 2)'.\Diamond$

The progress of QPSolver for the problem of Example 6.2 is shown in Figure 6.2. In that figure, we have used y_1 and y_2 to denote the coordinate axes and $x_0, x_1, \ldots, x_4$ to denote the iterates obtained by QPSolver.

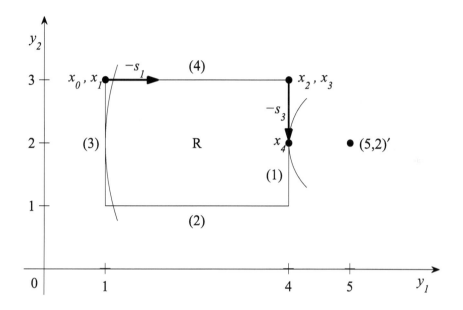

Figure 6.2 Geometry of Example 6.2.

The termination properties of QPSolver are formulated in the following theorem.

Theorem 6.1 *Let QPSolver be applied to the stated model problem and let Assumption 6.1 be satisfied. In addition, assume the model problem has no degenerate quasi-stationary points. Then the optimal solution for the model problem will be determined in a finite number of steps.*

Proof: Let $x_1, x_2, \ldots, x_j, x_{j+1}, \ldots$ be the iterates determined by the algorithm. By construction,

$$f(x_{j+1}) \le f(x_j), \quad j = 0, 1, \ldots .$$

The algorithm has the property that it will find a quasi-stationary point in at most n iterations. Let $j_0 < j_1 < \ldots < j_i, < \ldots$ denote the iterations for which $x_{j_0}, x_{j_1}, \ldots, x_{j_i}, \ldots$ are quasi-stationary points. For ease of notation, let ν denote the iteration number of a typical quasi-stationary point. Because termination has not yet occurred, $(u_\nu)_k < 0$, where k is determined by Step 3.2 as the index of the constraint being deleted from the active set. It now follows from Exercise 6.1(b) that

$$g'_\nu s_\nu > 0. \tag{6.25}$$

We next argue by contradiction that $\sigma_\nu > 0$. Suppose $\sigma_\nu = 0$. For simplicity, let k be the index of the constraint being dropped from the active set in Step 3.2 and let l be the index for the maximum feasible step size in Step 2. Furthermore, let $\tilde{A}'$ denote the matrix of gradients of active constraints specified by K_ν; i.e., all of the columns of A'_ν with the exception of a_k. Then because x_ν is a quasi-stationary point, we have

$$g_\nu = \tilde{A}' v_1 + w_1 a_k. \tag{6.26}$$

where v_1 and w_1 are the multipliers obtained in Step 1. Since A_{j+1} is obtained from A_j by replacing a_k with a_l and because $\sigma_\nu = 0$, $x_{\nu+1}$ is also a quasi-stationary point so

$$g_{\nu+1} = g_\nu = \tilde{A}' v_2 + w_2 a_l, \tag{6.27}$$

where v_2 and w_2 are the multipliers obtained in Step 1 at iteration $\nu + 1$. It now follows from (6.26) and (6.27) that

$$\tilde{A}'(v_1 - v_2) + w_1 a_k - w_2 a_l = 0. \tag{6.28}$$

From Step 3.2, $w_1 \neq 0$. Therefore, (6.28) shows that the gradients of those constraints active at x_ν are linearly dependent and thus x_ν is a degenerate quasi-stationary point. But this is precluded in the statement of the theorem. The assumption that $\sigma_\nu = 0$ leads to a contradiction and is therefore false. Consequently,

$$\sigma_\nu > 0. \tag{6.29}$$

From Taylor's theorem,

$$f(x_\nu - \sigma s_\nu) = f(x_\nu) - \sigma g'_\nu s_\nu + \frac{1}{2} \sigma^2 s'_\nu C s_\nu, \tag{6.30}$$

which with (6.25) implies that $f(x_\nu - \sigma s_\nu)$ is a strictly decreasing function of σ for all σ with $0 \leq \sigma \leq 1$. It now follows from (6.29) that $f(x_{\nu+1}) < f(x_\nu)$. Consequently,

$$f(x_{j_0}) > f(x_{j_1}) > \cdots > f(x_{j_i}) > \cdots$$

and no quasi-stationary point can ever be repeated. There are only finitely many quasi-stationary points and the optimal solution is one of them. Therefore QPSolver will terminate with the optimal solution in a finite number of steps.   □

In the development and presentation of QPSolver, we have assumed C is positive definite. This was made for simplicity and so we could focus on the most essential aspects of quadratic programming. The positive definite assumption can be weakened to positive semidefinite. This is done in Best [2], but is beyond the scope of this text. Being able to solve a QP with a positive semidefinite Hessian is important because covariance matrices in practice are sometimes positive semidefinite due to lack of data. A semidefinite capability would also allow the solution of linear programming problems with $\Sigma = 0$.

At each iteration of QPSolver, linear equations are required to be solved. We have purposefully left unspecified the method by which they are to be solved, although in the MATLAB code for QPSolver, we have solved linear equations by using the inverse matrix. We have used the inverse matrix for simplicity of presentation. When one wishes to develop a QP algorithm for a specific class of problems (for example portfolio optimization problems), it is best to utilize the structure of the problem and solve the linear equations by (hopefully) solving a smaller linear system. Some details of this are shown in Best [2] and van de Panne and Whinston [24].

Practitioners always like to solve large problems. Thus the computational complexity of QPSolver is of interest. To see where the majority of computations take place (or equivalently the computationally intensive parts of the algorithm), suppose at a typical iteration j, H_j has dimension (p, p). Inverting or factoring H_j from scratch requires approximately p^3 arithmetic operations. However, H_{j+1} differs from H_j by the addition or deletion of a single row and column and using an appropriate version of the Sherman and Morrison [23] and Woodbury [25] update formula, H_{j+1} can be determined from H_j in approximately p^2 operations, a considerable saving. Similar savings can be obtained using matrix factorizations.

Continuing the analysis with H_j being (p, p), solving the linear equations takes about p^2 computations. Computation of the maximum feasible step size takes about $2mn$ operations, although this could be smaller if some of the constraints were simple lower and upper bounds. All other computations require order n or m calculations. Consequently, solving linear equations with coefficient matrix H_j and updating H_j are the most computationally intensive parts of QPSolver.

It is shown in Best [2] that most active set quadratic programming algorithms construct the same sequence of points when applied to the same problem with the same starting point and when ties are resolved in the same manner. The total number of iterations is something that we cannot, in general, control. Therefore, to improve the efficiency of a QP algorithm it is most beneficial to look at ways that the computations involving H_j can be reduced. This is done for portfolio optimization problems in [2] where it is shown that all computations involving H_j for QPSolver (and its generalization to semidefinite C) can be performed by solving a much smaller linear system.

6.2   Computer Programs

We illustrate QPSolver by using it to solve the following problem.

Example 6.3

$$\begin{aligned} \text{minimize}: & \quad c'x + \tfrac{1}{2}x'Cx \\ \text{subject to}: & \quad -x_i \leq 0, \ i = 1, 2, \ldots, 5, \\ \text{and} & \quad x_1 + x_2 + x_3 + x_4 + x_5 = 1, \end{aligned}$$

where $m = 5$, $q = 1$, $c = (1, -2, 3, -4, 5)'$ and $C = BigC = \text{diag}(1, 2, 3, 4, 5)$. We take $x_0 = (0.2, 0.2, 0.2, 0.2, 0.2)'$ as the initial feasible point. ◇

The m-file Example6p3.m (see Figure 6.3) illustrates how QPSolver (Figure 6.5) is called to solve the problem of Example 6.3. Lines 2 to 11 set up the problem data. Line 12 calls checkdata (Figure 1.5) to test if C is positive definite. In line 13, the routine checkconstraints (Figure 6.4) is called to verify that the initial point x0 satisfies all of the problem constraints. Next, line 14 makes the call to QPSolver (Figure 6.5) to solve the given problem. Lines 15 to 22 print out the optimal solution xopt, the multipliers uj, the final index set Kj and the optimal objective function value fxj.

```
 1  %Example6p3.m
 2  A = [ −1 0 0 0 0; 0 −1 0 0 0; 0 0 −1 0 0 ; 0 0 0 −1 0 ; ...
 3        0 0 0 0 −1; 1 1 1 1 1 ]
 4  b = [ 0 0 0 0 0 1 ]'
 5  c = [ 1 −2 3 −4 5]'
 6  BigC = [1 0 0 0 0; 0 2 0 0 0 ; 0 0 3 0 0 ; 0 0 0 4 0 ; ...
 7         0 0 0 0 5 ]
 8  x0 = [0.2 0.2 0.2 0.2 0.2]'
 9  n = 5
10  m = 5
11  q = 1
12  checkdata(BigC,1.e−6);
13  checkconstraints(A,b,x0,m,q,n)
14  [minvalue,xopt,Kj,uj] = QPSolver(A,b,BigC,c,n,m,q,x0);
15
16  msg = 'optimal solution found'
17  xopt
18  msg = 'multipliers'
19  uj
20  msg = 'final index set'
21  Kj
22  msg = 'optimal objective function value'
23  minvalue
```

Figure 6.3   Example6p3.m.

The match up between the textual notation and the MATLAB notation used in the m-files presented in this section is shown in Table 6.1.

TABLE 6.1   **Match Up of Textual and MATLAB Notation**

Text Notation	MATLAB Notation
m	m
n	n
q	q
x_j	xj
C	BigC
A	A
b	b
x_0	x0
$f(x_j)$	fxj
g_j	gxj
s_j	sj
v_j	vj
u_j	uj
k	kay
K_j	Kj
σ_j	sigmaj
$\hat{\sigma}_j$	sigmajhat
l	ell
A_j	Aj
H_j	Hj

The function checkconstraints.m (Figure 6.4) tests the feasibility of the given initial feasible point x_0. Lines 4 and 5 check the feasibility of the inequality constraints (if any) at x0 and the indices of any violated constraints are added to the set infeas. Similarly, lines 8 and 9 check the feasibility of the equality constraints (if any) at x0. The indices of any violated equality constraints are added to the set infeas. Following these computations, if the size of infeas is zero, then x0 is indeed feasible and checkconstraints returns to the calling program. If the size of infeas is strictly positive, then x0 is infeasible and this is reported in lines 12 through 15. The indices of the violated constraints are output from line 13 and then the program is "paused" as in checkdata.m (Figure 1.5).

```
1   %checkconstraints.m
2   function checkconstraints(A,b,x0,m,q,n)
3   tol = 1.e-8;
4   if m > 0    % inequality constraints, if any
5      infeas = find(A(1:m,:)*x0 - b(1:m)>tol);
6   end
7   if q > 0    % equality constraints, if any
8      infeas = [infeas find(abs(A(m+1:m+q,:)*x0 ...
9                 -b(m+1:m+q))>tol)+m];
10  end
11  if size(infeas,1) > 0
12      errmsg = 'The following constraints are infeasible'
13      infeas
14      soln = 'Press Ctrl+C to continue'
15      pause
16  end
```

Figure 6.4   checkconstraints.m.

The function QPSolver is shown in Figure 6.5. Its input arguments are A, b, BigC, c, n, m, q and x0 which all have their usual meaning. The outputs are minvalue, minimizer, IndexSet and multipliers and these are defined in QPSolver.m. Initialization for the algorithm is performed in lines 25-39 as follows.

Line 23 defines the feasibility tolerance (tolfeas). If the left-hand side of constraint i evaluated at xo is within tolfeas of the right-hand side, then constraint i is considered to be active at x0. The variable flag is initialized to 0. It will be changed to 1 when the computations have detected an optimal solution. Line 26 constructs the initial index set (Kj) of inequality constraints active at x0. Line 27 calculates the number of constraints active at x0 from the size of the index set Kj. Line 28 constructs the matrix of gradients (transposed) of constraints active at x0. Lines 31–35 do the analogous thing for the equality constraints so that their gradients (transposed) are inserted into Aj below the gradients (transposed) of the active inequality constraints. Lines 37–39 initialize xj, gxj and fxj to the initial point x0, the gradient of f at x0 and the objective function value at x0, respectively. These three variables are global in scope to the scripts below QPSolver.

Lines 42–46 iteratively call the three steps of the QPSolver algorithm until flag has a nonzero value indicating an optimal solution has been found.

```
1   % QPSolver.m
2   function [minvalue,minimizer,IndexSet,multipliers] = ...
3       QPSolver(A,b,BigC,c,n,m,q,x0)
4   % INPUTS:
5   % A : linear constraint matrix
6   % b : right—hand side of constraints
7   % BigC : quadratic part of the objective function
8   % c : linear part of the objective function
9   % n : number of problem variables
10  % m : number of inequality constraints (first m rows of A)
11  % q : number of equality constraints (last q rows of A)
12  % x0 : initial point
13  % OUTPUTS:
14  % minvalue : minimum objective function value
15  % minimizer : the corresponding optimal solution
16  % IndexSet : the index set of constraints active at the
17  %               optimal solution
18  % uj : the constraint multipliers corresponding to the
19  %       optimal solution
20
21  % Initialization:
22
23  tolfeas = 1.e—8;  flag = 0;
24
25  if m > 0    % inequality constraints, if any
26      Kj = find(abs(A(1:m,:)*x0—b(1:m)) < tolfeas);
27      active = size(Kj,1);
28      Aj = A(Kj,:);
29  end
30
31  if q > 0    % equality constraints, if any
32      Kj = [Kj m+1:m+q];
33      active = active + q;
34      Aj = vertcat(Aj,A(m+1:m+q,:));
35  end
36
37  xj = x0;
38  gxj = c + BigC*xj;
39  fxj = c' * xj + 0.5 * xj' * BigC * xj;
40
```

```
41  while flag == 0
42      QPStep1
43      QPStep2
44      QPStep3
45  end
46
47  minvalue = fxj;
48  minimizer = xj;
49  IndexSet = Kj;
50  multipliers = uj;
```

Figure 6.5   QPSolver.m.

QPStep1.m (Figure 6.6) implements Step 1 of QPSolver. Line 3 constructs

$$\text{Hj} = \begin{bmatrix} \text{BigC} & \text{Aj}' \\ \text{Aj} & 0 \end{bmatrix},$$

and line 4 constructs

$$\text{rhs} = \begin{bmatrix} \text{gxj} \\ 0 \end{bmatrix}.$$

Line 5 solves the linear equations Hj y = rhs. Line 6 extracts the first n components into sj and line 7 extracts the last active components into vj.

```
1  %QPStep1.m
2  %Step 1: Computation of Search Direction sj.
3  Hj = vertcat([BigC Aj'],[Aj zeros(active,active)]);
4  rhs = vertcat(gxj, zeros(active,1));
5  y = Hj^-1 * rhs;
6  sj = y(1:n);
7  vj = y(n+1:n+active);
```

Figure 6.6   QPStep1.m.

QPStep2.m (Figure 6.7) implements Step 2 of QPSolver. Line 3 initializes ell (the index of the constraint giving the maximum feasible step size) to zero and tolstep (the step size tolerance) to a small number.

Line 5 constructs the set of indices of inactive constraints (inactive). Line 6 constructs sums which is a vector whose elements are $a_i's_j$ for all inactive constraints i (ordered according to inactive). Line 7 constructs ind which is a subset of inactive consisting of those i for which $a_i s_j <$ $-$tolstep. Using the sets ind and inactive, line 9 constructs the vector sigtest whose elements are

$$\frac{a_i' x_j - b_i}{a_i' s_j}$$

for all inactive constraints i with $a_i s_j < -$tolstep. Line 11 computes sigmajhat which is the smallest element of sigtest. Line 12 computes ell which is the index for which sigmajhat occurs. If the set ind has no elements, then line 14 sets sigmajhat to ∞. Line 17 sets sigmaj to the minimum of sigmajhat and 1.

```
1   %QPStep2.m
2   %Step 2: Computation of Step Size sigmaj.
3   ell = 0;   tolstep = 1.e-8;
4
5   inactive = setdiff(1:m,Kj);
6   sums = A(inactive,:)*sj;
7   ind=find(sums<-tolstep);
8   if ¬isempty(ind)
9       sigtest = (-b(inactive(ind))+A(inactive(ind),:)*xj)
10                 ... ./sums(ind);
11      sigmajhat = min(sigtest);
12      ell=inactive(ind(find(sigtest==sigmajhat,1)));
13  else
14      sigmajhat = Inf;
15  end;
16
17  sigmaj = min(sigmajhat,1);
```

Figure 6.7　　QPStep2.m.

QPStep3.m is shown in Figure 6.8. Lines 4, 5 and 6 update xj and compute the new values of the gradient and objective function, respectively. In line 8, if sigmaj $<$ 1 the function QStep3Sub1 is invoked with input arguments Kj, active, Aj, ell, A(ell,:) where the last argument gives row ell of A. The output quantities are the updated Kj, active and Aj. Otherwise (sigmaj = 1), the function QPStep3Sub2 is invoked with input arguments Kj, active, Aj, vj, m, and q which computes the updated versions of Kj, active, Aj, flag, and uj.

```
1   %QPStep3.m
2   %Step 3:   Update
3
4     xj = xj − sigmaj *sj;
5     gxj = c + BigC*xj;
6     fxj = c' * xj + 0.5 * xj' * BigC * xj;
7
8     if sigmaj < 1
9         [Kj,active,Aj]  = QPStep3Sub1(Kj,active,Aj,ell,A
10                          (ell,:));
11    else
12        [Kj,active,Aj,flag,uj] = QPStep3Sub2(Kj,active,Aj,vj,
13                          m,q);
14    end
```

Figure 6.8   QPStep3.m.

QPStep3Sub1.m which implements Step 3.1 of QPSolver, is shown in Figure 6.9. In line 6, the index set Kj_out is constructed by placing ell (the index of the new active constraint) in the first position followed by the elements of Kj. By the way, this is called in QPStep3, which effectively replaces Kj with the updated value. Line 7 computes active_out as the sum of active and 1. From the way this is called in QPStep3, this effectively replaces active with active + 1. Finally, line 8 computes Aj_out by vertically concatenating A_aug on top of Aj. From the way this called in QPStep3, Aj is replaced by adding row ell of A (the gradient (transposed) of the new active constraint with index ell) to the top row of Aj.

```
1   %QPStep3Sub1.m
2   function [Kj_out,active_out,Aj_out]= ...
3              QPStep3Sub1(Kj,active,Aj,ell,A_aug)
4
5   %Step 3.1:  New Active Constraint.
6        Kj_out = [ell Kj];
7        active_out = active + 1;
8        Aj_out = vertcat(A_aug,Aj);
```

Figure 6.9   QPStep3Sub1.m.

QPStep3Sub2.m, which implements Step 3.2 of QPSolver, is shown in Figure 6.10. The output arguments Kj_out, active_out, Aj_out, flag_out and uj_out contain the updated versions of Kj, active, Aj, flag and uj, respectively. From the way this is called in line 11 of QStep3.m (Figure 6.8), after the call to QPStep3Sub2.m Kj, active, Aj, flag and uj are all replaced by their updated values.

In lines 4 and 8 of QPStep3Sub2.m, Kj_out and Aj_out are set to their input values. Line 9 defines uj_out to be an m+q vector of zeroes. Line 10 constructs the components of uj_out corresponding to the multipliers for the inequality constraints using the index set Kj and the vector of multipliers − vj for the active inequality constraints obtained in Step 1. Note that the multipliers for the equality constraints are not computed in uj as they are not needed. Line 11 computes the smallest multiplier for the inequality constraints. If it is greater than or equal to − tolconv (tolconv is a small number defined on line 5), then we have an optimal solution. In this case, flag_out is set to 1 (indicating an optimal solution) and active_out is set to active. The vector uj_out is calculated as the full vector of multipliers, including those for the equality constraints. If the smallest multiplier for an inequality is less than tolconv, then control passes to line 20 which finds the index (pos) of the smallest multiplier (min(uj)) for an active inequality constraint. Lines 21 and 22 then construct Aj_out and Kj_out by removing row pos and element pos, respectively. Finally, active_out is obtained by reducing active by 1 and flag_out is set to zero.

```
1  function [Kj_out,active_out,Aj_out,flag_out,uj_out] = ...
2          QPStep3Sub2(Kj,active,Aj,vj,m,q)
3
4  Kj_out=Kj;
5  tolconv = 1.e-8;
6
7  % Step 3.2:  Drop an Active Constraint
8  Aj_out = Aj;
9  uj_out = zeros(1,m+q);
10 uj_out(Kj(1:active-q)) = -vj(1:active-q);
11 if min(uj_out) >= - tolconv
12 % Found optimal solution
13     flag_out = 1;
14     active_out = active;
15     uj_out = zeros(1,m+q);
16     uj_out(Kj(1:active)) = - vj(1:active);
17 else
18 % Want to drop constraint kay from position pos in Aj;
19 % Find pos
20     pos = find(min(uj_out) == -vj,1);
21     Aj_out(pos,:) = [];
22     Kj_out(pos) = [];
23     active_out = active - 1;
24     flag_out=0;
25 end
```

Figure 6.10   QPStep3Sub2.m.

6.3   Exercises

6.1 Let x_0 be feasible for (6.1) with active constraints $A_0 x_0 = b_0$. Assume A_0 has full row rank.

   (a) Show x_0 is a quasi-stationary point if and only if

$$\begin{bmatrix} C & A_0' \\ A_0 & 0 \end{bmatrix} \begin{bmatrix} s_0 \\ v_0 \end{bmatrix} = \begin{bmatrix} g_0 \\ 0 \end{bmatrix}$$

   has solution $s_0 = 0$.

(b) Now assume that x_1 is a quasi-stationary point for (6.1) and that A_1 is obtained from A_0 by deleting row k where $(v_0)_k > 0$. Show $g_1' s_1 > 0$ where s_1 is obtained from the solution of

$$\begin{bmatrix} C & A_1' \\ A_1 & 0 \end{bmatrix} \begin{bmatrix} s_1 \\ v_1 \end{bmatrix} = \begin{bmatrix} g_1 \\ 0 \end{bmatrix}.$$

6.2 Let x_0, s_0 and $f(x)$ be as in the discussion leading to (6.9). Show that $f(x_0 - \sigma s_0)$ is a decreasing function of σ for $\sigma \in [0, 1]$ and attains its minimum at $\sigma = 1$. (*Hint:* Use Lemma 6.1.)

6.3 (a) Apply QPSolver (by hand) to Example 5.2 using $x_0 = (1, 1)'$.

   (b) Verify your results for part (a) by using the QPSolver program as in Example 6.3 (Figure 6.3).

6.4 (a) Apply QPSolver (by hand) to Example 5.2 using $x_0 = (1, 0)'$.

   (b) Verify your results for part (a) by using the QPSolver program as in Example 6.3 (Figure 6.3).

6.5 Use the QPSolver program as in Example 6.3 (Figure 6.3) to solve the following problems with the indicated starting data. In each case, sketch the feasible region and show level sets for the objective function. In addition, state the multipliers for the optimality conditions.

(a)
$$\begin{array}{rrrrl}
\text{minimize}: & -4x_1 & -6x_2 & +x_1^2 + x_2^2 & \\
\text{subject to}: & x_1 & + x_2 & \le 2, & (1) \\
& -x_1 & - x_2 & \le 2, & (2) \\
& -x_1 & + x_2 & \le 2, & (3) \\
& x_1 & - x_2 & \le 2. & (4)
\end{array}$$

Initial data: $x_0 = (-2, 0)'$.

(b)
$$\begin{array}{rrrrl}
\text{minimize}: & -2x_1 & -2x_2 & +x_1^2 + x_2^2 - x_1 x_2 & \\
\text{subject to}: & -x_1 & + x_2 & \le -1, & (1) \\
& x_1 & + x_2 & \le 3, & (2) \\
& & -x_2 & \le 1. & (3)
\end{array}$$

Initial data: $x_0 = (3, -1)'$.

Chapter 7

Portfolio Optimization with Constraints

In all the theoretical developments so far, we have used the model problem

$$\min\{ - t\mu'x + \frac{1}{2}x'\Sigma x \mid l'x = 1\}, \tag{7.1}$$

or its extension to include a risk free asset. The parameter $t \geq 0$ quantifies the risk aversion of the investor. Although (7.1) is very useful for developing the basic concepts for portfolio optimization, it is not particularly suitable in practice. There are two main reasons for this. First, the solution of (7.1) may result in excessive *long* and *short selling*. An example of this is an optimal solution of (7.1) with $x_1 = 1000$, $x_2 = -1000$, $x_3 = 1$, $x_4 = 0, \ldots, x_n = 0$. This means that the investor would sell 1000 times his wealth in asset 2 in order to purchase 1000 times his wealth in asset 1. This is a completely unrealistic position. Furthermore, there are generally legal requirements restricting short sales. One way of precluding short sales is to impose nonnegativity restrictions $(x \geq 0)$ on the problem. Note that these constraints are a special case of the general linear inequality constraints considered in the previous chapter.

Another difficulty with (7.1) is that it may result in many very small trades, which may be mathematically correct, but would not be made because of transaction costs. See Best and Hlouskova [9] for a method to account for transaction costs.

The model problem we shall address in this chapter is

$$
\begin{aligned}
\text{minimize}: \quad & -t\mu'x + \tfrac{1}{2}x'\Sigma x \\
\text{subject to}: \quad & a_i'x \le b_i, \ i = 1, 2, \ldots, m, \\
& a_i'x = b_i, \ i = m+1, m+2, \ldots, m+q,
\end{aligned}
\tag{7.2}
$$

where the inequality constraints may include upper and lower bound constraints (including nonnegativity constraints as discussed above), sector constraints, the budget, and any other linear constraints. Sector constraints restrict asset holdings in particular areas. For example, if the investor's first 10 assets are all in oil related industries and the investor wishes to put no more than 5% in these, the linear constraints would include

$$
x_1 + x_2 + \cdots + x_{10} \le 0.05.
$$

We would like to solve (7.2) for all $t \ge 0$ and determine how the properties we have established for (7.1) generalize.

7.1 Linear Inequality Constraints: An Example

In this section, we begin our analysis of (7.2) by considering a small special case.

Example 7.1
We use problem (7.2) with data

$$
n = 3, \quad \mu = (1.05, 1.08, 1.1)', \quad \Sigma = diag(1 \times 10^{-2}, 2 \times 10^{-2}, 3 \times 10^{-2}),
$$

nonnegativity constraints ($x \ge 0$), and the budget constraint $x_1 + x_2 + x_3 = 1$. The nonnegativity constraints are greater-than-or-equal-to so we rewrite them as the equivalent $-x \leq 0$. Thus the constraints for this example are

$$
\begin{aligned}
-\ x_1 &\ \leq\ 0, &(1)\\
-\ x_2 &\ \leq\ 0, &(2)\\
-\ x_3 &\ \leq\ 0, &(3)\\
x_1\ +\ x_2\ +\ x_3 &\ =\ 1. &(4)
\end{aligned}
$$

For fixed t, Example 7.1 is just a quadratic programming problem and as such can be solved by QPSolver of Chapter 6. The MATLAB program "Example7p1.m," (see Figure 7.2), does precisely this for a variety of values of t. Note that in "Example7p1.m," the nonnegativity constraints $x \geq 0$ have been replaced by the equivalent $-x \leq 0$ since the model problem for QPSolver requires "$\leq$" constraints.

TABLE 7.1   Optimal Solutions with Various t
for Example 7.1

t	μ_p	σ_p^2	Optimal solution	Active Set
0.00	1.0673	0.0055	0.5455 0.2727 0.1818	4
0.11	1.0754	0.0063	0.3555 0.3427 0.3018	4
0.20	1.0820	0.0084	0.2000 0.4000 0.4000	4
0.31	1.0901	0.0125	0.0100 0.4700 0.5200	4
0.32	1.0906	0.0128	0.0000 0.4720 0.5280	4 , 1
0.40	1.0912	0.0133	0.0000 0.4400 0.5600	4 , 1
0.70	1.0936	0.0159	0.0000 0.3200 0.6800	4 , 1
1.49	1.0999	0.0298	0.0000 0.0040 0.9960	4 , 1
1.50	1.1000	0.0300	0.0000 0.0000 1.0000	4 , 1 , 2

Some of the results of running "Example7p1.m" are shown in Table 7.1. The full set of results can be obtained by running "Example7p1.m."

From Table 7.1, we can see that constraint 4 is active for all t in the table. This is as it should be, because the constraint with index 4 is the budget constraint, and it is an equality constraint, and is thus always active. Table 7.1 also shows that none of the nonnegativity constraints (indexed $1, 2, 3$ for $-x_1 \leq 0$, $-x_2 \leq 0$, and $-x_3 \leq 0$, respectively) are active for t between 0 and a little bit above 0.31. We therefore define the initial active set as K_0 as $K_0 = \{4\}$ for t slightly greater than 0.

Because $K_0 = \{4\}$, we define $A_0 = [1, 1, 1]$ and $b_0 = 1$. Then for t roughly in the interval $[0.0, 0.31]$, x_0 will be optimal for (7.1) (with the given problem data) if and only if it is optimal for

$$\min\{ -t\mu'x + \tfrac{1}{2}x'\Sigma x \mid A_0 x = b_0 \}. \tag{7.3}$$

According to Theorem 1.2, x_0 will be optimal for (7.3) if and only if x_0 together with the scalar multiplier for the budget constraint, v_0, satisfy the following linear equations

$$\begin{bmatrix} \Sigma & A_0' \\ A_0 & 0 \end{bmatrix} \begin{bmatrix} x_0 \\ v_0 \end{bmatrix} = \begin{bmatrix} t\mu \\ b_0 \end{bmatrix}. \tag{7.4}$$

In the right-hand side of (7.4), we have thought of t as a fixed scalar. In the present context, it is helpful to think of t as a scalar parameter which may vary in some interval. Rewriting (7.4) with the right-hand side separated into a constant vector and one depending on t gives

$$\begin{bmatrix} \Sigma & A_0' \\ A_0 & 0 \end{bmatrix} \begin{bmatrix} x_0 \\ v_0 \end{bmatrix} = \begin{bmatrix} 0 \\ b_0 \end{bmatrix} + t \begin{bmatrix} \mu \\ 0 \end{bmatrix}. \tag{7.5}$$

Let

$$H_0 = \begin{bmatrix} \Sigma & A_0' \\ A_0 & 0 \end{bmatrix}. \tag{7.6}$$

Then for the case at hand,

$$H_0 = \begin{bmatrix} 1 \times 10^{-2} & 0 & 0 & 1 \\ 0 & 2 \times 10^{-2} & 0 & 1 \\ 0 & 0 & 3 \times 10^{-2} & 1 \\ 1 & 1 & 1 & 0 \end{bmatrix}. \tag{7.7}$$

Note that by Lemma 6.2, H_0 is nonsingular. The solution of (7.5) is thus

$$\begin{bmatrix} x_0 \\ v_0 \end{bmatrix} = H_0^{-1} \begin{bmatrix} 0 \\ b_0 \end{bmatrix} + t H_0^{-1} \begin{bmatrix} \mu \\ 0 \end{bmatrix}$$

$$= \begin{bmatrix} h_{00} \\ g_{00} \end{bmatrix} + t \begin{bmatrix} h_{10} \\ g_{10} \end{bmatrix}$$

(7.8)

where

$$\begin{bmatrix} h_{00} \\ g_{00} \end{bmatrix} = H_0^{-1} \begin{bmatrix} 0 \\ b_0 \end{bmatrix} \quad \text{and} \quad \begin{bmatrix} h_{10} \\ g_{10} \end{bmatrix} = H_0^{-1} \begin{bmatrix} \mu \\ 0 \end{bmatrix}. \quad (7.9)$$

Using (7.9) and MATLAB, we obtain h_{00}, g_{00}, h_{10} and g_{10} so that the optimal solution to the given problem is

$$\begin{bmatrix} x_0 \\ v_0 \end{bmatrix} = \begin{bmatrix} x_0(t) \\ v_0(t) \end{bmatrix} = \begin{bmatrix} 0.5455 \\ 0.2727 \\ 0.1818 \\ -0.0055 \end{bmatrix} + t \begin{bmatrix} -1.7273 \\ 0.6364 \\ 1.0909 \\ 1.0673 \end{bmatrix}. \quad (7.10)$$

Thus,

$$x_0 = \begin{bmatrix} 0.5455 \\ 0.2727 \\ 0.1818 \end{bmatrix} + t \begin{bmatrix} -1.7273 \\ 0.6364 \\ 1.0909 \end{bmatrix}. \quad (7.11)$$

Similarly, the full vector of multipliers can be written as

$$u_0 = u_{00} + t u_{10},$$

where the components of u_{00} and u_{10} corresponding to inactive constraints (those whose indices are not in K_0) are all zero. The components of u_{00} and u_{10} corresponding to active constraints (those whose indices are in K_0) are obtained from v_0 and K_0. Because none of the first three constraints are active and the budget constraint (constraint 4) is active, from v_0 we obtain $(u_0)_4 = -0.0055 + t\,1.0673$ and the full vector of multipliers is

$$u_0 = \begin{bmatrix} 0.0 \\ 0.0 \\ 0.0 \\ -0.0055 \end{bmatrix} + t \begin{bmatrix} 0.0 \\ 0.0 \\ 0.0 \\ 1.0673 \end{bmatrix},$$

or, more simply

$$u_0 = (0, 0, 0, -0.0055 + t\, 1.0673)' \qquad (7.12)$$

Notice that in (7.11), both $(x_0)_2$ and $(x_0)_3$ are increasing functions of t so that as t is increased, their associated nonnegativity constraints will never become active. However, $(x_0)_1$ is a decreasing function of t and is reduced to zero when

$$t = t_1 = 0.5455/1.7273 = 0.3158. \qquad (7.13)$$

Defining $t_0 = 0$, we have shown that $x_0(t)$ is optimal for this example with multiplier vector $u_0(t)$ for all t with $t_0 \le t \le t_1$. Note that the only active constraint in this interval is the budget constraint. It is an equality constraint so that the optimality conditions impose no sign constraint on its multiplier (see Theorem 5.2).

We had previously observed from Table 7.1 that the active set changes from $K_0 = \{4\}$ to $K_1 = \{4, 1\}$ for t somewhere between 0.31 and 0.32. We have shown above that this change takes place at precisely $t_1 \equiv 0.3158$.

For t with $0 \le t \le t_1$, we can also calculate the efficient portfolio's expected return,

$$\mu_p = \mu' x_0(t) = \mu'(h_{00} + t h_{10}) = \alpha_{00} + \alpha_{10} t$$

where

$$\alpha_{00} = \mu' h_{00}, \quad \text{and} \quad \alpha_{10} = \mu' h_{10}.$$

We can also calculate the efficient portfolio's variance for t in this range;

$$\sigma_p^2 = x_0'(t)\Sigma x_0(t) = (h_{00} + t h_{10})'\Sigma(h_{00} + t h_{10}) = \beta_{00} + \beta_{10} t + \beta_{20} t^2$$

where $\beta_{00} = h_{00}'\Sigma h_{00}$, $\beta_{10} = 2h_{00}'\Sigma h_{10}$, and $\beta_{20} = h_{10}'\Sigma h_{10}$.

Note that each of the quantities h_{00}, h_{10}, α_{00}, α_{10}, β_{00}, β_{10}, and β_{20} are the analogs of h_0, h_1, α_0, α_1, β_0, β_1 and β_2, respectively, used in

Chapter 2. The second subscript of "0" indicates they correspond to interval 0 ($t_0 = 0 \leq t \leq t_1 = 0.3158$).

It is straightforward to show that

$$\beta_{10} = 0 \quad \text{and} \quad \alpha_{10} = \beta_{20}. \qquad (7.14)$$

The argument is identical for the model of Chapter 2. See Exercises 2.2, 2.3 and 7.3. Thus for this first interval, $\mu_p = 1.0673 + 0.0736t$ and $\sigma_p^2 = 0.0055 + 0.0736t^2$ for all t with $0 \leq t \leq t_1 = 0.3158$.

At the end of this first interval, x_1 is reduced to zero. This suggests that for $t > t_1$, constraints 1 ($-x_1 \leq 0$) and 4 (the budget constraint) will be active. Thus, we define $K_1 = \{4, 1\}$ and consequently

$$A_1 = \begin{bmatrix} 1 & 1 & 1 \\ -1 & 0 & 0 \end{bmatrix} \quad \text{and} \quad b_1 = \begin{bmatrix} 1 \\ 0 \end{bmatrix}.$$

Then for the next interval, (7.7) with A_0 replaced by A_1, becomes

$$H_1 = \begin{bmatrix} 1 \times 10^{-2} & 0 & 0 & 1 & -1 \\ 0 & 2 \times 10^{-2} & 0 & 1 & 0 \\ 0 & 0 & 3 \times 10^{-2} & 1 & 0 \\ 1 & 1 & 1 & 0 & 0 \\ -1 & 0 & 0 & 0 & 0 \end{bmatrix}. \qquad (7.15)$$

Note that by Lemma 6.2, H_1 is nonsingular. The solution of (7.5) with A_0 replaced by A_1, is thus

$$\begin{bmatrix} x_1 \\ v_1 \end{bmatrix} = H_1^{-1} \begin{bmatrix} 0 \\ b_1 \end{bmatrix} + t H_1^{-1} \begin{bmatrix} \mu \\ 0 \end{bmatrix},$$

$$= \begin{bmatrix} h_{01} \\ g_{01} \end{bmatrix} + t \begin{bmatrix} h_{11} \\ g_{11} \end{bmatrix} \qquad (7.16)$$

where

$$\begin{bmatrix} h_{01} \\ g_{01} \end{bmatrix} = H_1^{-1} \begin{bmatrix} 0 \\ b_1 \end{bmatrix} \quad \text{and} \quad \begin{bmatrix} h_{11} \\ g_{11} \end{bmatrix} = H_1^{-1} \begin{bmatrix} \mu \\ 0 \end{bmatrix}. \qquad (7.17)$$

Using (7.17) and MATLAB, we obtain h_{01}, g_{01}, h_{11} and g_{11} so that the optimal solution to the given problem is

$$
\begin{bmatrix} x_1 \\ v_1 \end{bmatrix} = \begin{bmatrix} x_1(t) \\ v_1(t) \end{bmatrix} = \begin{bmatrix} 0.0 \\ 0.6 \\ 0.4 \\ -0.012 \\ -0.012 \end{bmatrix} + t \begin{bmatrix} 0.0 \\ -0.4 \\ 0.4 \\ 1.088 \\ 0.0380 \end{bmatrix}. \tag{7.18}
$$

Thus,

$$
x_1 = \begin{bmatrix} 0.0 \\ 0.6 \\ 0.4 \end{bmatrix} + t \begin{bmatrix} 0.0 \\ -0.4 \\ 0.4 \end{bmatrix}. \tag{7.19}
$$

Similarly, the full vector for the multipliers can be written as

$$
u_1 = u_{01} + tu_{11},
$$

where the components of u_{01} and u_{11} corresponding to inactive constraints (those whose indices are not in K_1) are all zero. The components of u_{01} and u_{11} corresponding to active constraints (those whose indices are in K_1) are obtained from v_1 and K_1. Constraint 1 is active and we obtain its multiplier from v_1 as $(u_1)_1 = -0.012 + t0.038$. Similarly, from v_1 we obtain the multiplier for the budget constraint $(u_1)_4 = -0.012 + t\,1.088$. Because neither of the second or third non-negativity constraints is active, the full vector of multipliers is thus

$$
u_1 = \begin{bmatrix} -0.012 \\ 0.0 \\ 0.0 \\ -0.012 \end{bmatrix} + t \begin{bmatrix} 0.038 \\ 0.0 \\ 0.0 \\ 1.088 \end{bmatrix},
$$

or, more simply

$$
u_1 = (-0.012 + t0.038, 0, 0, -0.012 + t1.088)'. \tag{7.20}
$$

As t is increased beyond t_1, $(u_1)_1$ is an increasing function of t and thus imposes no restriction on t. Also, $(u_1)_4$ is the multiplier for the budget constraint and since the budget constraint is an equality constraint, there are no sign restrictions on it. However, from (7.19) $(x_1)_2$ is a decreasing function of t and is reduced to 0 when

$$t = t_2 = 0.6/0.4 = 1.5.$$

Thus we have shown that x_1 defined by (7.19) is optimal with multiplier vector u_1 given by (7.20) for all t in the interval $[t_1, t_2]$. Furthermore, it is easy to calculate the coefficients of the mean and variance for this interval as

$$\alpha_{01} = 1.088, \quad \alpha_{11} = 0.008, \quad \beta_{01} = 0.012, \quad \text{and} \quad \beta_{21} = 0.008.$$

When $t = t_2$, constraint 2 becomes active. In order to investigate the situation as t is increased from t_2, we define the next active set to be $\{4, 1, 2\}$. Consequently, we define

$$A_2 = \begin{bmatrix} 1 & 1 & 1 \\ -1 & 0 & 0 \\ 0 & -1 & 0 \end{bmatrix} \quad \text{and} \quad b_2 = \begin{bmatrix} 1 \\ 0 \\ 0 \end{bmatrix}.$$

Then for the next interval, (7.7) with A_0 replaced by A_2 becomes

$$H_2 = \begin{bmatrix} 1 \times 10^{-2} & 0 & 0 & 1 & -1 & 0 \\ 0 & 2 \times 10^{-2} & 0 & 1 & 0 & -1 \\ 0 & 0 & 3 \times 10^{-2} & 1 & 0 & 0 \\ 1 & 1 & 1 & 0 & 0 & 0 \\ -1 & 0 & 0 & 0 & 0 & 0 \\ 0 & -1 & 0 & 0 & 0 & 0 \end{bmatrix}. \quad (7.21)$$

The solution of (7.5) with A_0 replaced by A_2 is thus

$$\begin{bmatrix} x_2 \\ v_2 \end{bmatrix} = H_2^{-1} \begin{bmatrix} 0 \\ b_2 \end{bmatrix} + tH_2^{-1} \begin{bmatrix} \mu \\ 0 \end{bmatrix}$$

$$= \begin{bmatrix} h_{02} \\ g_{02} \end{bmatrix} + t \begin{bmatrix} h_{12} \\ g_{12} \end{bmatrix},$$

where

$$\begin{bmatrix} h_{02} \\ g_{02} \end{bmatrix} = H_2^{-1} \begin{bmatrix} 0 \\ b_2 \end{bmatrix} \quad \text{and} \quad \begin{bmatrix} h_{12} \\ g_{12} \end{bmatrix} = H_2^{-1} \begin{bmatrix} \mu \\ 0 \end{bmatrix}. \quad (7.22)$$

Using (7.22) and MATLAB, we obtain h_{02}, g_{02}, h_{12} and g_{12} so that the optimal solution to the given problem is

$$
\begin{bmatrix} x_2 \\ v_2 \end{bmatrix} = \begin{bmatrix} x_2(t) \\ v_2(t) \end{bmatrix} = \begin{bmatrix} 0.0 \\ 0.0 \\ 1.0 \\ -0.03 \\ -0.03 \\ -0.03 \end{bmatrix} + t \begin{bmatrix} 0.0 \\ 0.0 \\ 0.0 \\ 1.1 \\ 0.05 \\ 0.02 \end{bmatrix}. \tag{7.23}
$$

Thus,

$$
x_2 = \begin{bmatrix} 0.0 \\ 0.0 \\ 1.0 \end{bmatrix} + t \begin{bmatrix} 0.0 \\ 0.0 \\ 0.0 \end{bmatrix}. \tag{7.24}
$$

The vector of multipliers is

$$
u_2(t) = \begin{bmatrix} -0.03 \\ -0.03 \\ 0.0 \\ -0.03 \end{bmatrix} + t \begin{bmatrix} 0.05 \\ 0.02 \\ 0.0 \\ 1.1 \end{bmatrix} = u_{02} + t u_{12}. \tag{7.25}
$$

The coefficients of the mean and variance for this interval are readily calculated as

$$
\alpha_{02} = 1.1, \quad \alpha_{12} = 0.0, \quad \beta_{02} = 0.03, \quad \text{and} \quad \beta_{22} = 0.0.
$$

The multipliers for the first two nonnegativity constraints are both increasing functions of t. The multiplier for the budget constraint (an equality constraint) is unconstrained in sign. Furthermore, (7.24) shows that the optimal solution is precisely $(0, 0, 1)'$ which is independent of t and this point is optimal for all $t \geq t_2$. Note that this is in agreement with Exercise 5.7 and that asset 3 has the largest expected return of all three assets.

The results for Example 7.1 are summarized in Table 7.2. Note that $t_3 = \infty$.

Figure 7.1 shows the entire efficient frontier for Example 7.1. It consists of two pieces; one for the first interval $(0 = t_0 \leq t \leq t_1)$ and

TABLE 7.2   **Optimal, Piecewise Linear Solution for Example 7.1 and Associated Efficient Frontier Parameters**

j	0	1	2
t_j	0	0.3158	1.5
K_j	$\{4\}$	$\{4,1\}$	$\{4,1,2\}$

	h_{00}	h_{01}	h_{02}
	0.5455	0.0	0.0
h_{0j}	0.2727	0.6	0.0
	0.1818	0.4	1.0

	h_{10}	h_{11}	h_{12}
	-1.7273	0.0	0.0
h_{1j}	0.6364	-0.4	0.0
	1.0909	0.4	0.0

	u_{00}	u_{01}	u_{02}
	0.0	-0.012	-0.03
u_{0j}	0.0	0.0	-0.03
	0.0	0.0	0.0
	-0.0055	-0.012	-0.03

	u_{10}	u_{11}	u_{12}
	0.0	0.038	0.05
u_{1j}	0.0	0.0	0.02
	0.0	0.0	0.0
	1.0673	1.088	1.1

α_{0j}	1.0673	1.088	1.1
α_{1j}	0.0736	0.008	0.0
β_{0j}	0.0055	0.012	0.03
β_{2j}	0.0736	0.008	0.0

the second for the next interval ($t_1 \le t \le t_2$). The two "pieces" of the efficient frontier meet at $t = t_1$. Although the two pieces are defined by two different quadratic functions, the efficient frontier is, in fact, differentiable. To see why this is true, observe that in the first interval, $\mu_p = \alpha_{00} + \alpha_{10}t$ and $\sigma_p^2 = \beta_{00} + \beta_{20}t^2$. Then because $\alpha_{10} = \beta_{20}$, and by the chain rule of calculus,

$$\frac{d\mu_p}{d\sigma_p^2} = \frac{d\mu_p}{dt}\frac{dt}{d\sigma_p^2} = \frac{\alpha_{10}}{2t\beta_{20}} = \frac{1}{2t}.$$

The division by β_{20} is valid because $\beta_{20} \neq 0$. It is also true that

$$\frac{d\mu_p}{d\sigma_p^2} = = \frac{1}{2t} \tag{7.26}$$

for the second interval. The argument is precisely the same as for the first, noting that $\beta_{21} \neq 0$. The slope of the efficient frontier is thus $1/(2t)$ for all t with $0 < t < 1.5$. In particular, it is differentiable at $t = t_1$.

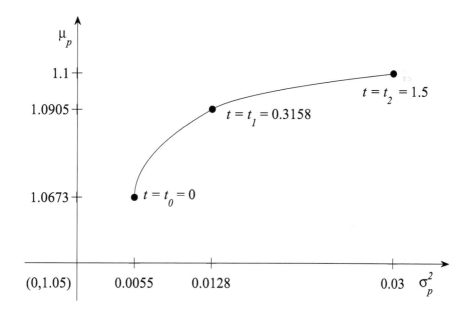

Figure 7.1   Two Pieces of the Efficient Frontier.

Although the computations for Example 7.1 were somewhat lengthy and perhaps even tedious, the essential logic was quite simple. Everything follows from the active set. The active set K_0 is obtained by solving (7.2) for $t = 0$. K_0 defines A_0 and thus H_0. The quantities in (7.9) are obtained in terms of H_0^{-1}. This gives us the optimal solution $x_0(t)$ and the associated multipliers $u_0(t)$ as linear functions of t. Next, an upper limit on t is obtained from the first inactive constraint to become active and the first multiplier to be reduced to zero. If the limit on t is determined by some previously inactive constraint becoming active, its index is added to K_0 to give K_1. If the limit on t corresponds to a multiplier being reduced to 0, the K_1 is obtained from K_0 by deleting the index of this constraint. The process continues with the new active set.

Note that Example 7.1 has a diagonal covariance matrix and the constraints are nonnegativity constraints on all assets plus the budget constraint. Also notice that in the minimum variance portfolio, all assets are held positively. Then as t is increased, the asset with the smallest expected return was reduced to zero and always remains at zero as t is increased further. As t is further increased, the asset with the second smallest expected return is reduced to zero and remains there. Then the asset with the third smallest expected return is reduced to zero and stays there. Best and Hlouskova [7] have shown that this same result will hold for problems of n assets with a similar covariance matrix and constraint structure.

7.2   The General Case

In this section, we consider the model problem

$$
\begin{aligned}
\text{minimize}: \quad & -t\mu'x + \tfrac{1}{2}x'\Sigma x \\
\text{subject to}: \quad & a_i'x \leq b_i, \ i = 1, 2, \ldots, m, \\
& a_i'x = b_i, \ i = m+1, m+2, \ldots, m+q,
\end{aligned}
\tag{7.27}
$$

and generalize the results of Example 7.1 to this problem.

Let t^* be such that $t^* \geq 0$. Let QPSolver be applied to (7.27) and for some integer j (to be discussed later and not to be confused with the iteration counter for QPSolver), let H_j, A_j, b_j and K_j be the final quantities found by the algorithm when it obtains an optimal solution. Then K_j contains an ordered set of the indices of the active constraints, the rows of A_j contain the transposes of the gradients of the active constraints, ordered according to K_j and b_j contains the corresponding right-hand sides. Then

$$H_j = \begin{bmatrix} C & A'_j \\ A_j & 0 \end{bmatrix}.$$

According to Theorem 1.2, x_j will be optimal for (7.27) if and only if x_j together with the multiplier vector for the active constraints, v_j, satisfy the following linear equations

$$\begin{bmatrix} \Sigma & A'_j \\ A_j & 0 \end{bmatrix} \begin{bmatrix} x_j \\ v_j \end{bmatrix} = \begin{bmatrix} t^*\mu \\ b_j \end{bmatrix}, \qquad (7.28)$$

nonnegativity of those components of v_j corresponding to inequality constraints, and x_j must satisfy those constraints of (7.27) which are not in the active set K_j. If t^* in (7.28) is replaced by a parameter t, where t is allowed to vary in some neighborhood of t^*, then (7.28) can be written

$$\begin{bmatrix} \Sigma & A'_j \\ A_j & 0 \end{bmatrix} \begin{bmatrix} x_j \\ v_j \end{bmatrix} = \begin{bmatrix} 0 \\ b_j \end{bmatrix} + t \begin{bmatrix} \mu \\ 0 \end{bmatrix}. \qquad (7.29)$$

Since H_j constructed by QPSolver is always nonsingular, we can write the solution of (7.29) as

$$\begin{bmatrix} x_j \\ v_j \end{bmatrix} = H_j^{-1} \begin{bmatrix} 0 \\ b_j \end{bmatrix} + t H_j^{-1} \begin{bmatrix} \mu \\ 0 \end{bmatrix},$$

$$= \begin{bmatrix} h_{0j} \\ g_{0j} \end{bmatrix} + t \begin{bmatrix} h_{1j} \\ g_{1j} \end{bmatrix}, \qquad (7.30)$$

where

$$\begin{bmatrix} h_{0j} \\ g_{0j} \end{bmatrix} = H_j^{-1} \begin{bmatrix} 0 \\ b_j \end{bmatrix} \quad \text{and} \quad \begin{bmatrix} h_{1j} \\ g_{1j} \end{bmatrix} = H_j^{-1} \begin{bmatrix} \mu \\ 0 \end{bmatrix}. \qquad (7.31)$$

Now we have the optimal portfolio holdings $x_j(t)$ and the multipliers for the active linear constraints, $v_j(t)$, as

$$x_j(t) = h_{0j} + t h_{1j} \quad \text{and} \quad v_j(t) = g_{0j} + t g_{1j}. \qquad (7.32)$$

We next want to determine restrictions on t such that the components $v_j(t)$ corresponding to active inequality constraints remain nonnegative. We also want to determine restrictions on t such that $x_j(t)$ satisfies those inequality constraints of (7.27) which are inactive when $t = t^*$. We need to identify which of those components of $v_j(t)$ correspond to inequality constraints and which correspond to equality constraints. To this end, we introduce the ordered index $K_j = \{k_1, k_2, \ldots, k_{n_j}\}$, where

$$A'_j = [a_{k_1}, a_{k_2}, \ldots, a_{k_{n_j}}]. \tag{7.33}$$

For example, in Example 7.1, $K_0 = \{4\}$ and $A'_0 = [a_4]$ for the first interval and $K_1 = \{4, 1\}$ and $A'_1 = [a_4, a_1]$ for the second interval.

We require that $(v_j)_i(t) \geq 0$ where $v_j(t)$ is defined in (7.32) and i is such that constraint k_i is an active inequality constraint. Using (7.32), this leads to the requirements that for all i with $1 \leq k_i \leq m$,

$$t \leq -\frac{(g_{0j})_i}{(g_{1j})_i} \text{ for } (g_{1j})_i < 0, \tag{7.34}$$

and

$$t \geq -\frac{(g_{0j})_i}{(g_{1j})_i} \text{ for } (g_{1j})_i > 0. \tag{7.35}$$

Inequality (7.34) must be satisfied for all i with $1 \leq k_i \leq m$ and $(g_{1j})_i < 0$. The largest value of t, compatible with all these upper bounds is the smallest upper bound. Thus by defining

$$\tilde{t}_{ju} = \min\{ -\frac{(g_{0j})_i}{(g_{1j})_i} \mid (g_{1j})_i < 0 \text{ and } 1 \leq k_i \leq m \},$$

it follows that

$$(v_j)_i(t) \geq 0 \text{ for all } t \text{ with } 1 \leq k_i \leq m \text{ and } t^* \leq t \leq \tilde{t}_{ju}. \tag{7.36}$$

Inequality (7.35) imposes lower bounds on t. Proceeding in an analogous manner as in obtaining $\tilde{t}_{ju}$, we define

$$\tilde{t}_{jl} = \max\{ -\frac{(g_{0j})_i}{(g_{1j})_i} \mid (g_{1j})_i > 0 \text{ and } 1 \leq k_i \leq m \}.$$

It now follows that

$$(v_j)_i(t) \geq 0 \text{ for all } t \text{ with } \tilde{t}_{jl} \leq t \leq \tilde{t}_{ju}, \text{ and } i \text{ with } 1 \leq k_i \leq m. \quad (7.37)$$

Another way of stating (7.37) is that the multipliers for the active inequality constraints remain nonnegative for all t in the interval $[\tilde{t}_{jl}, \tilde{t}_{ju}]$. Note that we have used "l" and "u" as part of the subscripts of $\tilde{t}_{jl}$ and $\tilde{t}_{ju}$ to indicate lower and upper bounds, respectively.

We next turn to restrictions on t imposed by inequality constraints which are inactive at t^*. Let $x_j(t)$ be as in (7.32). Then $x_j(t)$ must satisfy

$$a_i' x_j(t) \leq b_i \text{ for all with } i \notin K_j \text{ and } 1 \leq i \leq m. \quad (7.38)$$

Substitution of the expression for $x_j(t)$ in (7.32) into (7.38) gives the following restrictions. For all i with $1 \leq i \leq m$ and $i \notin K_j$,

$$t \leq \frac{b_i - a_i' h_{0j}}{a_i' h_{1j}} \text{ for } a_i' h_{1j} > 0, \quad (7.39)$$

and

$$t \geq \frac{b_i - a_i' h_{0j}}{a_i' h_{1j}} \text{ for } a_i' h_{1j} < 0. \quad (7.40)$$

Inequality (7.39) must be satisfied for all i with $i \notin K_j$ and $1 \leq i \leq m$. The largest value of t, compatible with all these upper bounds, is the smallest upper bound. Thus by defining

$$\hat{t}_{ju} = \min\{ \frac{b_i - a_i' h_{0j}}{a_i' h_{1j}} \mid a_i' h_{1j} > 0, i \notin K_j \text{ and } 1 \leq i \leq m \}, (7.41)$$

it follows that

$$a_i' x_j(t) \leq b_i, \text{ for all } t \text{ with } t^* \leq t \leq \hat{t}_{ju} \text{ and } i = 1, 2, \ldots, m. \quad (7.42)$$

Inequality (7.40) imposes lower bounds on t. Proceeding in an analogous manner as in obtaining $\hat{t}_{ju}$, we define

$$\hat{t}_{jl} = \max\{ \frac{b_i - a_i' h_{0j}}{a_i' h_{1j}} \mid a_i' h_{1j} < 0, i \notin, K_j \text{ and } 1 \leq i \leq m \}. (7.43)$$

It now follows that

$x_j(t)$ satisfies all constraints of (7.27) for all t with $\hat{t}_{jl} \leq t \leq \hat{t}_{ju}$.(7.44)

Define

$$t_j \;=\; \max\{\tilde{t}_{jl}, \hat{t}_{jl}\} \text{ and } t_{j+1} \;=\; \min\{\tilde{t}_{ju}, \hat{t}_{ju}\}. \tag{7.45}$$

Any point t^* on the efficient frontier for (7.27) corresponds to an interval $[t_j, t_{j+1}]$ and t^* lies in this interval. The optimal solution $x_j(t)$ and the associated vector $v_j(t)$ are given by (7.32). All such quantities are uniquely determined by the set of active constraints specified by K_j. There are only finitely many such active sets and so there are only finitely many intervals. The end of each interval must also be the start of some other interval, for if not, there would be a point between two intervals which does not belong to any interval. But then our analysis has shown that this point is contained in some interval. Thus, for any interval $[t_j, t_{j+1}]$, there must be a second interval $[t_k, t_{k+1}]$ with $t_{j+1} = t_k$.

Thus, we can reorder the indices of the intervals such that they satisfy

$$[t_0, t_1], \; [t_1, t_2], \ldots, [t_{j-1}, t_j], \; [t_j, t_{j+1}], \ldots, [t_{N-1}, t_N].$$

We have used N to denote the number of such intervals. Other than a few special cases, it is not possible to determine N *a priori*. However, it is easy to see that

$$N \leq 2^m,$$

because 2^m is the number of subsets of $\{1, 2, \ldots, m\}$; i.e., the maximum number of active sets for (7.27).

We next turn to the shape of the efficient frontier. Consider t in the j-th interval for some j with $0 \leq N - 1$ and t with $t_j \leq t \leq t_{j+1}$. It follows from (7.32) that the efficient portfolios in this interval can be written as

$$x_j(t) \;=\; h_{0j} + th_{1j}.$$

From this, it is straightforward to calculate the portfolio's mean, μ_{pj}, as

$$\mu_{pj} = \alpha_{0j} + t\alpha_{1j}, \tag{7.46}$$

where

$$\alpha_{0j} = \mu' h_{0j} \quad \text{and} \quad \alpha_{1j} = \mu' h_{1j}. \tag{7.47}$$

Similarly, the portfolio's variance, σ_{pj}^2, is

$$\sigma_{pj}^2 = \beta_{0j} + \beta_{1j} t + \beta_{2j} t^2, \tag{7.48}$$

where

$$\beta_{0j} = h'_{0j} \Sigma h_{0j}, \quad \beta_{1j} = 2h'_{0j} \Sigma h_{1j} \quad \text{and} \quad \beta_{2j} = h'_{1j} \Sigma h_{1j}. \tag{7.49}$$

Analogous to the case of just the budget constraint in Chapter 2, these coefficients satisfy

$$\beta_{2j} = \alpha_{1j} \quad \text{and} \quad \beta_{1j} = 0. \tag{7.50}$$

See Exercises 2.2, 2.3 and 7.3.

Similar to (2.16), we solve for t in (7.46) and (7.48) and then eliminate t to give

$$\sigma_{pj}^2 - \beta_{0j} = (\mu_{pj} - \alpha_{0j})^2 / \beta_{2j}. \tag{7.51}$$

Equation (7.51) gives the equation of the j-th segment of the efficient frontier for (7.27). It is valid provided

$$\beta_{2j} \neq 0, \tag{7.52}$$

and when (7.52) is satisfied, the corresponding efficient frontier segment is defined by (7.46) and (7.48) for all $t \in [t_j, t_{j+1}]$ or equivalently, all μ_p and σ_p^2 satisfying

$$\alpha_{0j} + t_j \alpha_{1j} \leq \mu_p \leq \alpha_{0j} + t_{j+1} \alpha_{1j} \tag{7.53}$$

and

$$\beta_{0j} + t_j^2 \beta_{2j} \leq \sigma_p^2 \leq \beta_{0j} + t_{j+1}^2 \beta_{2j}. \tag{7.54}$$

Figure 7.1 shows the two pieces of the efficient frontier for Example 7.1. The end of the first interval and the start of the next interval occurs for t_1. Although the curvature of the efficient frontier does change here (compare the extension of the first interval to the second interval), it is not clear from Figure 7.1 whether or not it is differentiable at t_1. It turns out that it usually (but not always) is differentiable.

To investigate this further, we calculate the slope of the efficient frontier in the j-th interval $[t_j, t_{j+1}]$. Using (7.46), (7.48) and the chain rule of calculus,

$$\frac{d\mu_p}{d\sigma_p^2} = \frac{d\mu_p}{dt}\frac{dt}{d\sigma_p^2} = \frac{\alpha_{1j}}{2t\beta_{2j}} = \frac{1}{2t}, \tag{7.55}$$

provided

$$\beta_{2j} \neq 0. \tag{7.56}$$

The result that the slope of the efficient frontier is always $1/(2t)$ (provided (7.56) is satisfied), is quite surprising because the slope does not depend on the problem. It implies the efficient frontier for any portfolio optimization problem has the identical slope.

Note that the restriction (7.56) has already shown up in (7.52).

The following theorem summarizes the results of this section.

Theorem 7.1 *The efficient frontier for (7.27) is characterized by a finite number of intervals* $[t_j, t_{j+1}]$, $j = 0, 1, \ldots, N - 1$ *with* $t_0 = 0$ *and* $t_j < t_{j+1}$, $j = 0, 1, \ldots, N - 1$. *In each parametric interval,* $t_j \leq t \leq t_{j+1}$, $j = 0, 1, \ldots, N - 1$, *the following properties hold.*

(a) The efficient portfolios are given by (7.32).

(b) The expected return μ_{pj} and the variance σ_{pj}^2 of these portfolios are given by (7.46) and (7.48), respectively, where the coefficients are given by (7.47) and (7.49) and satisfy (7.50). The corresponding segment of the efficient frontier is given by (7.51).

(c) If $\beta_{2j} \neq 0$ for $j = 0, 1, \ldots, N - 1$, then the entire efficient frontier is differentiable and has slope $\frac{d\mu_p}{d\sigma_p^2} = \frac{1}{2t}$.

Theorem 7.1(a) states that the efficient portfolios and the associated multipliers for the active constraints are piecewise linear functions of the risk aversion parameter t.

The portfolios

$$\hat{x}_0 \equiv h_{00}, \text{ together with } \hat{x}_{j+1} \equiv (h_{0j} + t_{j+1}h_{1j}), \ j = 0, 1, \ldots, N - 1 \tag{7.57}$$

are called the *corner portfolios*. With $t_0, t_1, \ldots, t_N$ they can be used to construct all of the efficient portfolios for the given problem.

Although Table 7.2 does give the entire efficient frontier, it is sometimes informative to view the corner portfolios, their expected returns and variances. For Example 7.1, the corner portfolios are shown in Table 7.3.

TABLE 7.3   Corner Portfolios for Example 7.1

j	0	1	2
t_j	0	0.3158	1.5
K_j	$\{4\}$	$\{4, 1\}$	$\{4, 1, 2\}$
$\mu_p(t_j)$	1.0673	1.0905	1.1
$\sigma_p^2(t_j)$	0.0055	0.0128	0.03
	$\hat{x}_0$	$\hat{x}_1$	$\hat{x}_2$
	0.5455	0.0	0.0
$\hat{x}_j$	0.2727	0.4737	0.0
	0.1818	0.5263	1.0

Note that $x_j(t_{j+1}) = x_{j+1}(t_{j+1})$ even though they are given by different formulae. This is because they are both optimal solutions for (7.27) for $t = t_{j+1}$ and from Theorem 5.3 the optimal solution is uniquely determined.

7.3   Computer Results

Figure 7.2 shows the program Example7p1.m which was used to construct the values given in Table 7.1. Lines 2–9 set up the problem data for Example 7.1 and check it with checkdata and checkconstraints.

```
1   %Example7p1.m
2   A = [ −1 0 0; 0 −1 0; 0 0 −1; 1 1 1 ];
3   b = [ 0 0 0 1 ]';
4   mu = [ 1.05 1.08 1.1 ]';
5   BigC = [1 0 0; 0 2 0; 0 0 3]/100.;
6   n = 3;   m = 3;   q = 1;
7   x0 = [0.2 0.2 0.6 ]';
8   checkdata(BigC,1.e−6);
9   checkconstraints(A,b,x0,m,q,n);
10  tlow = .0;   tinc = 0.01;   thigh = 2.0
11  ii = 0;
12  for t = tlow:tinc:thigh;
13     c = − t * mu;
14     [minvalue,xj,Kj,uj] = QPSolver(A,b,BigC,c,n,m,q,x0);
15     t
16     xjprint = xj'
17     mup = mu' * xj
18     if t == 0.
19        mupmin = mup;
20     end
21     sigma2p = xj' * BigC * xj
22     ii = ii + 1;
23     sig2vec(ii) = sigma2p;
24     mupvec(ii) = mup;
25     muplowvec(ii) = 2*mupmin − mup;
26     Kj
27  end
28
29  plot(sig2vec,mupvec,'−k',sig2vec,muplowvec,'−−k')
30  xlabel('Portfolio Variance \sigma_p^2')
31  ylabel('Portfolio Mean \mu_p')
```

Figure 7.2   Example 7.1: Example7p1.m.

Lines 10 to 12 set up a loop which goes from $t = 0$ to $t = 2$ in increments of 0.01. The counter ii is initialized to 0. The linear part of the objective function for QPSolver is set in line 13 and QPSolver is called in line 14. The risk aversion parameter t, the efficient portfolio xprint and the expected return are printed lines 15 through 17. The variance, expected return and lower point on the efficient frontier are stored in the vectors sig2vec, mupvec and muplowvec, in lines 23 through 25. Next, the active set Kj obtained from QPSolver is printed. Finally, the information stored in the three vectors is plotted. Axes labels are defined in lines 30 and 31.

Figure 7.3 shows the program Example7p1plot.m which plots the two pieces of the efficient frontier for Example 7.1. The data defining the two intervals ($\alpha_{00} = $ alpha00, $\alpha_{01} = $ alpha01, etc.) is taken from Table 7.3 and is defined in lines 1 to 4. The mean and variance is calculated for points in the first interval and stored in the vectors muvec and varvec in lines 5 to 12. These values are plotted in line 13 and lines

```
1   %Example7p1plot.m
2   alpha00 = 1.0673; alpha10 = 0.0736; alpha01 = 1.088;
3   alpha11 = 0.008;
4   beta00 = 0.0055; beta20 = 0.0736; beta01 = 0.0122;
5   beta21 = 0.008;
6   t0 = 0.0; t1 = 0.3158; t2 = 1.5;
7   tinc = (t1 - t0)/100.;
8   i = 1; muvec = []; varvec = [];
9   for t = t0:tinc:t1
10      muvec(i) = alpha00 + t*alpha10;
11      varvec(i) = beta00 + (t^2)*beta20;
12      i = i + 1;
13  end
14  plot(varvec,muvec,'-k')   %plot first interval; t0-t1
15  text(varvec(1)+0.001,muvec(1),'\leftarrow t_0')
16  text(varvec(101)+0.001,muvec(101),'\leftarrow t_1')
17  text(varvec(1),muvec(1),'\bullet')
18  text(varvec(101),muvec(101),'\bullet')
19
20  hold on
21  tmax = 0.6;
22  tinc = (tmax - t1)/100.;
```

```
23  i = 1;   muvec = []; varvec = [];
24  for t = t1:tinc:tmax
25      muvec(i) = alpha00 + t*alpha10;
26      varvec(i) = beta00 + (t^2)*beta20;
27      i = i + 1;
28  end
29
30  %plot extension of first interval t1->t2
31  plot(varvec,muvec,'--k')
32
33  hold on
34  tinc = (t2 - t1)/100.;
35  i = 1;   muvec = []; varvec = [];
36  for t = t1:tinc:t2
37      muvec(i) = alpha01 + t*alpha11;
38      varvec(i) = beta01 + (t^2)*beta21;
39      i = i + 1;
40  end
41  plot(varvec,muvec,'-k')
42  text(varvec(101),muvec(101),'\bullet')
43  text(varvec(101)+0.001,muvec(101),'\leftarrow t_2')
44  text(0.007,1.075, ...
45          '\leftarrow {\rm efficient frontier, first
46                      piece}')
47  text(0.02,1.095,  ...
48          '\leftarrow {\rm efficient frontier, second
49                      piece}')
50  text(0.0095,1.103, ...
51          ' {\rm extension of first piece} \rightarrow')
52  xlabel('\sigma_p^2')
53  ylabel('\mu_p')
54  hold on
55  tmax = t1;
56  tinc = (tmax - t0)/100.;
57  i = 1;   muvec = []; varvec = [];
58  for t = t0:tinc:tmax
59      muvec(i) = alpha01 + t*alpha11;
60      varvec(i) = beta01 + (t^2)*beta21;
61      i = i + 1;
62  end
63
64  plot(varvec,muvec,'--k')
```

Figure 7.3   Example 7.1: Example7p1plot.m.

14 to 17 add some helpful annotations. Lines 20 to 27 plot the extension of the first piece of the efficient frontier using dashes. Here, "extension" means points of the form $(\beta_{00} + t^2\,\beta_{02},\ \alpha_{00} + t\,\alpha_{01}\,t)$ for $t \geq t_1$.

Lines 32 to 40 plot the second piece of the efficient frontier and lines 41 to 50 add some descriptive information. Finally, lines 51 to 61 plot the extension of the second piece of the efficient frontier. This turns out to be quite small and shows up as a single dash to the left of t_1.

The graph obtained from running Example7p1plot is shown in Figure 7.4.

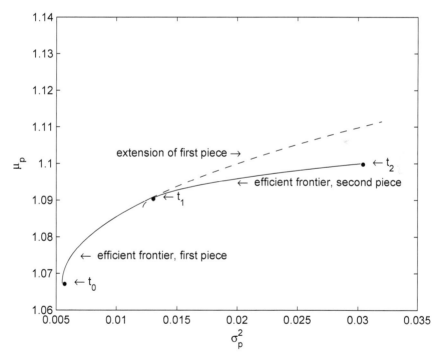

Figure 7.4   Graph from Example7p1plot.m.

7.4   Exercises

7.1 Continue Example 7.1 by observing that for $t = t_2$, constraints 1, 2 and 4 are active. Construct A_2 and H_2 accordingly and verify that $h_{12} = 0$ so that the optimal solution is $(0, 0, 1)'$ for all $t \geq t_2$.

7.2 Apply Theorem 7.1 to the risk free asset problem

$$\text{minimize}: \quad -t\,(\mu', r) \begin{bmatrix} x \\ x_{n+1} \end{bmatrix} + \tfrac{1}{2} \begin{bmatrix} x \\ x_{n+1} \end{bmatrix}' \begin{bmatrix} \Sigma & 0 \\ 0' & 0 \end{bmatrix} \begin{bmatrix} x \\ x_{n+1} \end{bmatrix}$$

$$\text{subject to}: \qquad l'x + x_{n+1} = 1, \quad x_{n+1} \geq 0.$$

Note that this problem has an explicit nonnegativity constraint on x_{n+1}. Beginning with $t_0 = 0$, find the corner portfolios and the equation of the efficient frontier.

7.3 Verify (equation 7.50); i.e., verify that $\beta_{2j} = \alpha_{1j}$ and $\beta_{1j} = 0$ for all $j = 0, 1, \ldots$.

7.4 Show that any efficient portfolio for (7.2) can be written as

$$\theta \hat{x}_j + (1 - \theta)\hat{x}_{j+1}$$

for some j with $0 \leq j \leq N - 1$ and for some scalar θ satisfying $0 \leq \theta \leq 1$. This shows that the entire set of efficient portfolios can be generated by the corner portfolios.

Chapter 8

Determination of the Entire Efficient Frontier

8.1 The Entire Efficient Frontier

The model problem we shall address in this chapter is

$$
\begin{aligned}
\text{minimize}: \quad & -t\mu'x + \tfrac{1}{2}x'\Sigma x \\
\text{subject to}: \quad & a_i'x \leq b_i, \ i = 1, 2, \ldots, m, \\
& a_i'x = b_i, \ i = m+1, m+2, \ldots, m+q,
\end{aligned}
\tag{8.1}
$$

which is identical to the model problem (7.2) used in Chapter 7. Theorem 7.1 showed that the optimal solution for (8.1), was a piecewise linear function of the risk aversion parameter t. The result was an existence result and did not tell us how to construct the optimal solution. It is the purpose of this chapter to formulate an algorithm for the solution of (8.1) for all $t \geq 0$. Throughout this chapter, we assume that Σ is positive definite.

The problem (8.1) is a special case of a parametric quadratic programming problem (PQP). The more general problem not only allows for the parameter to be in the linear part of the objective function, but

165

also in the right-hand side of the constraints. In addition, Σ is allowed to be positive semidefinite. The method we present here (we call it PQPSolver) is a special case of that presented in Best [8].

Let $t_0 = 0$ and suppose the QP obtained from (8.1) by replacing t with t_0 has been solved using QPSolver giving the optimal active set K_0 and corresponding matrix of gradients of active constraints A_0. Now suppose for some integer $j \geq 0$ we have the optimal solution x_j for (8.1) with t replaced by t_j. Let the optimal active set be K_j and corresponding matrix of gradients of active constraints A_j and let $\hat{b}_j$ be the corresponding vector of right-hand sides. Let v_j be the multipliers for the active constraints. We have these quantities for $j = 0$. We could obtain them for $j \geq 1$ using QPSolver, but it turns out that this is unnecessary because we can obtain K_{j+1} from K_j in an iterative manner.

Let

$$H_j = \begin{bmatrix} \Sigma & A'_j \\ A_j & 0 \end{bmatrix}.$$

Since we have an optimal solution for (8.1) for $t = t_j$, according to Theorem 1.2, x_j will be optimal for (8.1) if and only if (1) x_j together with the multiplier vector for the active constraints, v_j, satisfy the linear equations

$$\begin{bmatrix} \Sigma & A'_j \\ A_j & 0 \end{bmatrix} \begin{bmatrix} x_j \\ v_j \end{bmatrix} = \begin{bmatrix} t_j \mu \\ \hat{b}_j \end{bmatrix}, \tag{8.2}$$

(2) those components of v_j corresponding to inequality constraints are nonnegative, and (3) x_j must satisfy those constraints of (8.1) which are not in the active set K_j. If t_j in (8.2) is replaced by a parameter t, where t is allowed to increase above t_j, then (8.2) can be written

$$\begin{bmatrix} \Sigma & A'_j \\ A_j & 0 \end{bmatrix} \begin{bmatrix} x_j \\ v_j \end{bmatrix} = \begin{bmatrix} 0 \\ \hat{b}_j \end{bmatrix} + t \begin{bmatrix} \mu \\ 0 \end{bmatrix}. \tag{8.3}$$

Since H_j constructed by QPSolver is always nonsingular, we can write the solution of (8.3) as

$$
\begin{bmatrix} x_j \\ v_j \end{bmatrix} = H_j^{-1} \begin{bmatrix} 0 \\ \hat{b}_j \end{bmatrix} + t H_j^{-1} \begin{bmatrix} \mu \\ 0 \end{bmatrix},
$$

$$
= \begin{bmatrix} h_{0j} \\ g_{0j} \end{bmatrix} + t \begin{bmatrix} h_{1j} \\ g_{1j} \end{bmatrix},
$$

(8.4)

where

$$
\begin{bmatrix} h_{0j} \\ g_{0j} \end{bmatrix} = H_j^{-1} \begin{bmatrix} 0 \\ \hat{b}_j \end{bmatrix} \quad \text{and} \quad \begin{bmatrix} h_{1j} \\ g_{1j} \end{bmatrix} = H_j^{-1} \begin{bmatrix} \mu \\ 0 \end{bmatrix}.
$$

(8.5)

Now we have the optimal portfolio holdings $x_j(t)$ and the multipliers for the active linear constraints, $v_j(t)$, as

$$
x_j(t) = h_{0j} + t h_{1j} \quad \text{and} \quad v_j(t) = g_{0j} + t g_{1j} .
$$

(8.6)

We next need to determine restrictions on t such that the components of $v_j(t)$ corresponding to active inequality constraints remain nonnegative. We also need to determine restrictions on t such that $x_j(t)$ satisfies those inequality constraints of (8.1) which are inactive when $t = t_j$. We need to identify which of those components of $v_j(t)$ correspond to inequality constraints and which correspond to equality constraints. To this end, we introduce the ordered index $K_j = \{k_1, k_2, \ldots, k_{n_j}\}$, where

$$
A'_j = [a_{k_1}, a_{k_2}, \ldots, a_{k_{n_j}}] \quad \text{and} \quad \hat{b}_j = (b_{k_1}, b_{k_2}, \ldots, b_{k_{n_j}})'.
$$

We first address the question of nonnegativity of those multipliers for active inequality constraints. These constraints are precisely those which satisfy $1 \le k_i \le m$. If i satisfies this last condition then from (8.6) the multiplier for constraint k_i is

$$
(g_{0j})_i + (g_{1j})_i \, t.
$$

(8.7)

This last will be a decreasing function of t provided $(g_{1j})_i < 0$ and in this case we require

$$
t \le -\frac{(g_{0j})_i}{(g_{1j})_i} \quad \text{for all } i \text{ with } (g_{1j})_i < 0.
$$

(8.8)

The restriction imposed by (8.8) applies to all constraints in the active set $(1 \leq i \leq n_j)$ and which are inequality constraints $(1 \leq k_i \leq m)$. Define

$$I_j = \{\, i \mid 1 \leq i \leq n_j \text{ and } 1 \leq k_i \leq m \,\}.$$

Equation (8.8) imposes upper bounds on t. The largest value of t compatible with all of these bounds is

$$\tilde{t}_{j+1} = \min\{-\frac{(g_{0j})_i}{(g_{1j})_i} \mid \text{all } i \text{ with } (g_{1j})_i < 0, \text{ and } i \in I_j \,\} \quad (8.9)$$

$$= -\frac{(g_{0j})_{l_1}}{(g_{1j})_{l_1}}. \quad (8.10)$$

In (8.10), l_1 is the index for which the minimum occurs and for $t = \tilde{t}_{j+1}$ the multiplier for constraint k_{l_1} is reduced to zero.

If $(g_{1j})_i \geq 0$ for all i with $1 \leq i \leq n_j$ and $1 \leq k_i \leq m$, then all the multipliers for active inequality constraints are increasing functions of t and increasing t will impose no limits on t. We indicate this possibility by setting $\tilde{t}_{j+1} = \infty$.

It now follows that

$$(v_j)_i(t) \geq 0 \quad \text{for all } t \text{ with } 1 \leq k_i \leq m \text{ and } t_j \leq t \leq \tilde{t}_{j+1}. \quad (8.11)$$

The inequality constraints which are inactive at t_j will also impose restrictions on t and we now consider these. Let $x_j(t)$ be as in (8.6). Then $x_j(t)$ must satisfy

$$a_i' x_j(t) \leq b_i \quad \text{for all with } i \notin K_j \text{ and } 1 \leq i \leq m. \quad (8.12)$$

Using the expression for $x_j(t)$ in (8.6) into (8.12) gives the following restrictions. For all i with $1 \leq i \leq m$ and $i \notin K_j$,

$$t \leq \frac{b_i - a_i' h_{0j}}{a_i' h_{1j}} \quad \text{for } a_i' h_{1j} > 0. \quad (8.13)$$

Inequality (8.13) must be satisfied for all i with $i \notin K_j$ and $1 \leq i \leq m$. The largest value of t compatible with all these upper bounds is the smallest upper bound. Thus by defining

$$\hat{t}_{j+1} \;=\; \min\{ \frac{b_i - a_i' h_{0j}}{a_i' h_{1j}} \;\mid\; a_i' h_{1j} > 0,\; i \notin K_j \text{ and } 1 \le i \le m \,\},$$

$$\tag{8.14}$$

$$= \; \frac{b_{l_2} - a_{l_2}' h_{0j}}{a_{l_2}' h_{1j}}. \tag{8.15}$$

In (8.15), l_2 is the index for which the minimum occurs and for $t = \hat{t}_{j+1}$ constraint l_2 becomes active.

If $a_i' h_{1j} \le 0$ for all i with $i \notin K_j$ and $1 \le i \le m$, then no inactive inequality constraint becomes active as t is increased beyond t_j. We indicate this possibility by setting $\hat{t}_{j+1} = \infty$.

It now follows that

$$a_i' x_j(t) \le b_i, \quad \text{for all } t \text{ with } t_j \le t \le \hat{t}_{j+1} \text{ and } 1 \le i \le m,\; i \notin K_j.$$

$$\tag{8.16}$$

If we now define

$$t_{j+1} \;=\; \min\{ \tilde{t}_{j+1},\, \hat{t}_{j+1} \}, \tag{8.17}$$

then for all t with $t_j \le t \le t_{j+1}$ it follows that

$$(v_j)_i(t) \ge 0 \quad \text{for all } t \text{ with } 1 \le k_i \le m \tag{8.18}$$

and

$$a_i' x_j(t) \le b_i, \quad \text{for all } i \text{ with } 1 \le i \le m \text{ and } i \notin K_j. \tag{8.19}$$

Equations (8.18) and (8.19) show that $x_j(t)$ defined by (8.6) is optimal for (8.1) for all t with $t_j \le t \le t_{j+1}$.

In order to continue and find the optimal solution for the next interval $[t_{j+1}, t_{j+2}]$, we must modify the active set K_j to obtain K_{j+1}. How this is done depends on whether $t_{j+1} = \tilde{t}_{j+1}$ or $t_{j+1} = \hat{t}_{j+1}$.

If $t_{j+1} = \tilde{t}_{j+1}$, and $\tilde{t}_{j+1} < \infty$ let k_{l_1} be defined by (8.10). Then the multiplier for constraint k_{l_1} is reduced to zero at $t = t_{j+1}$ and would become strictly negative if t were allowed to increase beyond t_{j+1}. In this case we delete k_{l_1} from the active set K_j to obtain the new active set K_{j+1}. That is, we set $K_{j+1} = K_j - \{k_{l_1}\}$.

If $t_{j+1} = \hat{t}_{j+1}$ and $\hat{t}_{j+1} < \infty$, let l_2 be defined by (8.15). Then inequality constraint l_2, which was inactive at t_j, becomes active when $t = t_{j+1}$. If t were allowed to increase beyond $\hat{t}_{j+1}$, then inequality constraint i_2 would become violated. In this case, we add l_2 to K_j to form the new active K_{j+1}. That is, we set $K_{j+1} = K_j + \{l_2\}$.

If $t_{j+1} = \infty$, then we have the entire efficient frontier for all t with $0 \le t \le \infty$ and the algorithm is complete.

A detailed statement of the algorithm now follows.

PQPSolver

Model Problem:

$$\begin{aligned}
\text{minimize}: \quad & -t\mu'x + \tfrac{1}{2}x'\Sigma x \\
\text{subject to}: \quad & a_i'x \le b_i, \ i = 1, 2, \ldots, m, \\
& a_i'x = b_i, \ i = m+1, m+2, \ldots, m+q.
\end{aligned} \tag{8.20}$$

Initialization:

Let $t_0 = 0$ and solve the model problem with $t = t_0$ using QPSolver. Let $K_0 = \{k_1, k_2, \ldots k_{n_0}\}$, A_0 and $\hat{b}_0$ be the final index set, constraint matrix and corresponding right-hand side so obtained. Set $j = 0$ and go to Step 1.

Step 1:   Computation of Optimal Solution and Multipliers

Let

$$H_j = \begin{bmatrix} \Sigma & A_j' \\ A_j & 0 \end{bmatrix} \quad \text{and} \quad K_j = \{k_1, k_2, \ldots, k_{n_j}\}.$$

Compute the vectors h_{0j}, h_{1j}, g_{0j} and g_{1j} from

$$\begin{bmatrix} h_{0j} \\ g_{0j} \end{bmatrix} = H_j^{-1} \begin{bmatrix} 0 \\ \hat{b}_j \end{bmatrix} \quad \text{and} \quad \begin{bmatrix} h_{1j} \\ g_{1j} \end{bmatrix} = H_j^{-1} \begin{bmatrix} \mu \\ 0 \end{bmatrix},$$

and go to Step 2.

Step 2:   Computation of End of Interval

If there is at least one i with $(g_{1j})_i < 0$, $1 \leq i \leq n_j$ and $1 \leq k_i \leq m$, compute l_1 and $\tilde{t}_{j+1}$ according to

$$\tilde{t}_{j+1} = \min\left\{ -\frac{(g_{0j})_i}{(g_{1j})_i} \;\middle|\; \text{all } i \text{ with } (g_{1j})_i < 0, \ 1 \leq i \leq n_j \text{ and } 1 \leq k_i \leq m \right\}$$

$$= -\frac{(g_{0j})_{l_1}}{(g_{1j})_{l_1}}.$$

Otherwise, set $\tilde{t}_{j+1} = \infty$.

If there is at least one i with $i \notin K_j$ and $1 \leq i \leq m$, compute l_2 and $\hat{t}_{j+1}$ according to

$$\hat{t}_{j+1} = \min\left\{ \frac{b_i - a_i' h_{0j}}{a_i' h_{1j}} \;\middle|\; a_i' h_{1j} > 0, \ i \notin K_j \text{ and } 1 \leq i \leq m \right\},$$

$$= \frac{b_{l_2} - a_{l_2}' h_{0j}}{a_{l_2}' h_{1j}}.$$

Otherwise, set $\hat{t}_{j+1} = \infty$.

Set $t_{j+1} = \min\{\tilde{t}_{j+1}, \ \hat{t}_{j+1}\}$ and go to Step 3.

Step 3:   Update

Print "The optimal solution is $h_{0j} + t h_{1j}$ and the multipliers for the active constraints are $g_{0j} + t g_{1j}$ for all t in the interval $[t_j, t_{j+1}]$."

If $t_{j+1} = \infty$, print "The entire efficient frontier has been determined" and stop. Otherwise, if $t_{j+1} = \tilde{t}_{j+1}$, set $K_{j+1} = K_j - \{k_{l_1}\}$

and $n_{j+1} = n_j - 1$, and, if $t_{j+1} = \hat{t}_{j+1}$, set $K_{j+1} = K_j + \{l_2\}$ and $n_{j+1} = n_j + 1$.

Replace j with $j + 1$ and go to Step 1.

We illustrate the steps of PQPSolver by applying it to the problem of Example 7.1 and determine the entire efficient frontier.

Example 8.1

We repeat the data for Example 7.1. We use problem (8.1) with data

$$n = 3, \quad \mu = (1.05, 1.08, 1.1)', \quad \Sigma = diag(1 \times 10^{-2}, 2 \times 10^{-2}, 3 \times 10^{-2}),$$

nonnegativity constraints $(x \geq 0)$, and the budget constraint $x_1 + x_2 + x_3 = 1$. The nonnegativity constraints are greater than or equal to, so we rewrite them as the equivalent $-x \leq 0$. Thus the constraints for this example are

$$
\begin{array}{rcccccl}
- & x_1 & & & & \leq & 0, \quad (1) \\
& & - & x_2 & & \leq & 0, \quad (2) \\
& & & - & x_3 & \leq & 0, \quad (3) \\
& x_1 & + & x_2 & + & x_3 & = & 1 \quad (4)
\end{array}
$$

and we are to solve this special case of (8.1) for all $t \geq 0$.

The steps of PQPSolver are as follows.

Initialization:

Solving the problem for $t_0 = 0$ with QPSolver gives:

$K_0 = \{4\}, \quad A_0 = [1\ 1\ 1], \quad \hat{b}_0 = [1]$ and $n_0 = 1$. Set $j = 0$ and go to Step 1.

Iteration 0

Step 1:

$$
H_0 = \begin{bmatrix}
1 \times 10^{-2} & 0 & 0 & 1 \\
0 & 2 \times 10^{-2} & 0 & 1 \\
0 & 0 & 3 \times 10^{-2} & 1 \\
1 & 1 & 1 & 0
\end{bmatrix}, \quad \hat{b}_0 = [1],
$$

$$K_0 = \{4\}.$$

Compute the vectors h_{00}, h_{10}, g_{00} and g_{10}:

$$\begin{bmatrix} h_{00} \\ g_{00} \end{bmatrix} = H_0^{-1} \begin{bmatrix} 0 \\ \hat{b}_0 \end{bmatrix} = \begin{bmatrix} 0.5455 \\ 0.2727 \\ 0.1818 \\ -0.0055 \end{bmatrix}$$

and

$$\begin{bmatrix} h_{10} \\ g_{10} \end{bmatrix} = H_0^{-1} \begin{bmatrix} \mu \\ 0 \end{bmatrix} = \begin{bmatrix} -1.7273 \\ 0.6364 \\ 1.0909 \\ 1.0673 \end{bmatrix}.$$

Thus,

$$h_{00} = \begin{bmatrix} 0.5455 \\ 0.2727 \\ 0.1818 \end{bmatrix} \quad h_{10} = \begin{bmatrix} -1.7273 \\ 0.6364 \\ 1.0909 \end{bmatrix},$$

$$g_{00} = [-0.0055] \quad \text{and} \quad g_{10} = [1.0673].$$

Go to Step 2.

Step 2:
There is no i with $(g_{1j})_i < 0$ and $1 \le k_i \le m$. Set $\tilde{t}_1 = \infty$.

$$\begin{aligned} \hat{t}_1 &= \min \left\{ \frac{0.5455}{1.7273}, -, - \right\} = 0.3158, \\ &= \frac{b_1 - a_1' h_{00}}{a_1' h_{10}}, \quad l_2 = 1. \end{aligned}$$

Note that in the calculation of $\hat{t}_1$ we have used a dash $(-)$ to indicate those i with $i \notin K_0$ and $1 \le i \le 3$.

Set $t_1 = \min\{\infty, 0.3158\} = 0.3158$ and go to Step 3.

Step 3:
The optimal solution is

$$\begin{bmatrix} 0.5455 \\ 0.2727 \\ 0.1818 \end{bmatrix} + t \begin{bmatrix} -1.7273 \\ 0.6364 \\ 1.0909 \end{bmatrix}.$$

with multiplier $-0.0055 + t1.0673$ for constraint 4, for all $t \in [0, 0.3158]$.

Since $t_1 = \hat{t}_1$, set $K_1 = \{1, 4\}$, $n_1 = 2$

Set $j = 1$ and go to Step 1.

Iteration 1

Step 1:

$$
H_1 = \begin{bmatrix} 1 \times 10^{-2} & 0 & 0 & -1 & 1 \\ 0 & 2 \times 10^{-2} & 0 & 0 & 1 \\ 0 & 0 & 3 \times 10^{-2} & 0 & 1 \\ -1 & 0 & 0 & 0 & 0 \\ 1 & 1 & 1 & 0 & 0 \end{bmatrix}, \quad \hat{b}_1 = \begin{bmatrix} 0.0 \\ 1.0 \end{bmatrix},
$$

$$
K_1 = \{1, 4\}.
$$

Compute the vectors h_{01}, h_{11}, g_{01} and g_{11}:

$$
\begin{bmatrix} h_{01} \\ g_{01} \end{bmatrix} = H_1^{-1} \begin{bmatrix} 0 \\ \hat{b}_1 \end{bmatrix} = \begin{bmatrix} 0.0 \\ 0.6 \\ 0.4 \\ -0.012 \\ -0.012 \end{bmatrix}
$$

and

$$
\begin{bmatrix} h_{11} \\ g_{11} \end{bmatrix} = H_1^{-1} \begin{bmatrix} \mu \\ 0 \end{bmatrix} = \begin{bmatrix} 0.0 \\ -0.4 \\ 0.4 \\ 0.0380 \\ 1.088 \end{bmatrix}.
$$

Thus,

$$
h_{01} = \begin{bmatrix} 0.0 \\ 0.6 \\ 0.4 \end{bmatrix} \quad h_{10} = \begin{bmatrix} 0.0 \\ -0.4 \\ 0.4 \end{bmatrix},
$$

$$
g_{01} = \begin{bmatrix} -0.012 \\ -0.012 \end{bmatrix} \quad \text{and} \quad g_{11} = \begin{bmatrix} 0.0380 \\ 1.088 \end{bmatrix}.
$$

Note that because the ordering of the index set $K_1 = \{1, 4\}$, the components of g_{01} and g_{11} give the multipliers for constraints 1 and 4, in that order.

Go to Step 2.

Step 2:
There is no i with $(g_{11})_i < 0$ and $1 \le k_i \le m$. Set $\tilde{t}_2 = \infty$.

$$
\hat{t}_2 \;=\; \min\left\{-,\, \frac{0.6}{0.4},\, -\right\} = 1.50,
$$

$$
= \frac{b_2 - a_2' h_{01}}{a_2' h_{11}}, \quad l_2 = 2.
$$

Set $t_2 = \min\{\infty, 1.5\} = 1.5$ and go to Step 3.

Step 3:
The optimal solution is

$$
\begin{bmatrix} 0.0 \\ 0.6 \\ 0.4 \end{bmatrix} + t \begin{bmatrix} 0.0 \\ -0.4 \\ 0.4 \end{bmatrix},
$$

with full multiplier vector $(-0.012 + t\,0.038, 0, 0, -0.012 + t\,1.088)'$, for all $t \in [0.3158,\ 1.5]$.

Since $t_2 = \hat{t}_2$, set $K_2 = \{1, 2, 4\}$, $n_2 = 3$

Set $j = 2$ and go to Step 1.

Iteration 2

Step 1:

$$
H_2 = \begin{bmatrix}
1 \times 10^{-2} & 0 & 0 & -1 & 0 & 1 \\
0 & 2 \times 10^{-2} & 0 & 0 & -1 & 1 \\
0 & 0 & 3 \times 10^{-2} & 0 & 0 & 1 \\
-1 & 0 & 0 & 0 & 0 & 0 \\
0 & -1 & 0 & 0 & 0 & 0 \\
1 & 1 & 1 & 0 & 0 & 0
\end{bmatrix}, \quad \hat{b}_2 = \begin{bmatrix} 0.0 \\ 0.0 \\ 1.0 \end{bmatrix},
$$

$$
K_2 = \{1, 2, 4\}.
$$

Compute the vectors h_{02}, h_{12}, g_{02} and g_{12}:

$$\begin{bmatrix} h_{02} \\ g_{02} \end{bmatrix} = H_2^{-1} \begin{bmatrix} 0 \\ \hat{b}_2 \end{bmatrix} = \begin{bmatrix} 0.0 \\ 0.0 \\ 1.0 \\ -0.03 \\ -0.03 \\ -0.03 \end{bmatrix}$$

and

$$\begin{bmatrix} h_{12} \\ g_{12} \end{bmatrix} = H_2^{-1} \begin{bmatrix} \mu \\ 0 \end{bmatrix} = \begin{bmatrix} 0.0 \\ 0.0 \\ 0.0 \\ 0.05 \\ 0.02 \\ 1.1 \end{bmatrix}.$$

Thus,

$$h_{02} = \begin{bmatrix} 0.0 \\ 0.0 \\ 1.0 \end{bmatrix} \quad h_{12} = \begin{bmatrix} 0.0 \\ 0.0 \\ 0.0 \end{bmatrix},$$

$$g_{02} = \begin{bmatrix} -0.03 \\ -0.03 \\ -0.03 \end{bmatrix} \quad \text{and} \quad g_{12} = \begin{bmatrix} 0.05 \\ 0.02 \\ 1.1 \end{bmatrix}.$$

Note that because the ordering of the index set $K_2 = \{1, 2, 4\}$, the components of g_{02} and g_{12} give the multipliers for constraints 1, 2 and 4, in that order.

Go to Step 2.

Step 2:
There is no i with $(g_{12})_i < 0$ and $1 \leq k_i \leq m$. Set $\tilde{t}_2 = \infty$.

There is no i with $i \notin K_2$ and $1 \leq i \leq 3$. Set $\hat{t}_3 = \infty$.

Set $t_3 = \min\{\infty, \infty\} = \infty$ and go to Step 3.

Step 3:
The optimal solution is

$$
\begin{bmatrix} 0.0 \\ 0.0 \\ 1.0 \end{bmatrix} + t \begin{bmatrix} 0.0 \\ 0.0 \\ 0.0 \end{bmatrix},
$$

with full multiplier vector

$$
(-0.03, -0.03, 0.0, -0.03)' + t\,(0.05, 0.02, 0.0, 1.1)',
$$

for all $t \in [1.5, \infty)$. ◇

If one compares the results of Example 7.1 with the results of Example 8.1, the optimal solutions so obtained are identical. However, sometimes the computations appear to be different. For example in Example 7.1, we used

$$
A_1 = \begin{bmatrix} 1 & 1 & 1 \\ -1 & 0 & 0 \end{bmatrix}, \quad \hat{b}_1 = \begin{bmatrix} 1 \\ 0 \end{bmatrix}, \quad \text{and } K_1 = \{4, 1\}.
$$

In Example 8.1, we used

$$
A_1 = \begin{bmatrix} -1 & 0 & 0 \\ 1 & 1 & 1 \end{bmatrix}, \quad \hat{b}_1 = \begin{bmatrix} 0 \\ 1 \end{bmatrix}, \quad \text{and } K_1 = \{1, 4\}.
$$

The difference being that the rows of one A_1 are in the reverse order of the other. So too are the elements of K_1 and the components of b_1. The order that constraints are put in A_j does not matter as long as the components of b_j and the elements of K_j are defined in a consistent manner.

One might expect that this section would end with a theorem saying that PQPSolver will solve (8.1) in a finite number of steps. However, such a result requires a certain assumption. It could happen that $\tilde{t}_{j+1}$, $\hat{t}_{j+1}$ or t_{j+1} is not uniquely determined; i.e., there are ties in the computation of any of these quantities. In this case, the next active set K_{j+1} is not uniquely determined. If an arbitrary choice is made, it may lead to $t_{j+2} = t_{j+1}$; i.e., no progress is being made. Additional active sets could be tried, all making no progress.

This problem has been addressed by a number of authors. See Arseneau and Best [1] and Wang [26] for example. These authors give analytic solutions which require solving a derived QP. See Exercise 8.3 for an example of a problem where this difficulty occurs naturally. Perold [21] gives an algorithm for the solution of a PQP but does not take into account the problems caused by ties.

Another way of resolving the problem is to solve the QP (8.1) for fixed $t = t_{j+1} + \epsilon$ where ϵ is some small positive number. The final active set obtained in solving this with QPSolver should be the correct active set to continue PQPSolver. The problem is how to choose ϵ. If it is not small enough, a parametric interval may be missed.

However, with the assumption of no ties, PQPSolver will indeed determine the entire efficient frontier for (8.1) as formulated in the following theorem.

Theorem 8.1 *Let PQPSolver be applied to (8.1), let $t_0 = 0$ and let t_{j+1}, h_{0j}, h_{1j}, $j = 0, 1, \ldots, N-1$, be the results so obtained. For $j = 0, 1, \ldots, N-1$, assume that $\tilde{t}_{j+1}$, $\hat{t}_{j+1}$ and t_{j+1} are uniquely determined. Then for $j = 0, 1, \ldots, N-1$,*

(a) $t_j < t_{j+1}$,

(b) $x_j(t) = h_{0j} + t h_{1j}$ *are efficient portfolios for all t with $t_j \leq t \leq t_{j+1}$,*

(c) *the expected return of $x_j(t)$ is $\mu_{pj}(t) = \alpha_{0j} + \alpha_{1j} t$ where $\alpha_{0j} = \mu' h_{0j}$ and $\alpha_{1j} = \mu' h_{1j}$ for all t with $t_j \leq t \leq t_{j+1}$,*

(d) *and the variance of $x_j(t)$ is $\sigma_{pj}^2 = \beta_{0j} + \beta_{2j} t^2$ where $\beta_{0j} = h_{0j}' \Sigma h_{0j}$ and $\beta_{2j} = h_{1j}' \Sigma h_{1j}$ for all t with $t_j \leq t \leq t_{j+1}$.*

(e) *Each A_j has full row rank.*

Proof:
Under the stated assumption, part (a) is proven in Best [8]. The remainder follows by construction. □

There is a relationship between the coefficients α_{0j}, α_{1j}, β_{0j} and β_{2j} for adjacent intervals $[t_j,\ t_{j+1}]$ and $[t_{j+1},\ t_{j+2}]$ which we now proceed to develop. In what follows, all quantities are to be as in the statement of Theorem 8.1. From Theorem 5.3, the optimal solution for (8.1) is uniquely determined for each $t \geq 0$. Thus the optimal expected return at the end of the interval $[t_j,\ t_{j+1}]$ must be identical with that for the beginning of the interval $[t_{j+1},\ t_{j+2}]$. This implies

$$\alpha_{0j} + t_{j+1}\alpha_{1j} = \alpha_{0,j+1} + t_{j+1}\alpha_{1,j+1}.$$

Rewriting this gives

$$\alpha_{0j} - \alpha_{0,j+1} = t_{j+1}(\alpha_{1,j+1} - \alpha_{1j}). \tag{8.21}$$

Similarly for the variance at t_{j+1}:

$$\beta_{0j} + t_{j+1}^2\beta_{2j} = \beta_{0,j+1} + t_{j+1}^2\beta_{2,j+1},$$

which implies

$$\beta_{0j} - \beta_{0,j+1} = t_{j+1}^2(\beta_{2,j+1} - \beta_{2j}). \tag{8.22}$$

Let $w_j(t)$ denote the optimal objective function value for (8.1) for the j-th parametric interval $[t_j,\ t_{j+1}]$, for $j = 0, 1, \ldots, N - 1$. Because $\mu'x_j(t) = \alpha_{0j} + \alpha_{1j}t$ and $x_j(t)'\Sigma x_j(t) = \beta_{0j} + \beta_{2j}t^2$ for $t_j \leq t \leq t_{j+1}$, and $\alpha_{1j} = \beta_{2j}$ it follows that

$$
\begin{aligned}
w_j(t) &= -t(\alpha_{0j} + \alpha_{1j}t) + \frac{1}{2}(\beta_{0j} + \beta_{2j}t^2) \\
&= \frac{\beta_{0j}}{2} - \alpha_{0j}t + \left(-\alpha_{1j} + \frac{\beta_{2j}}{2}\right)t^2 \\
&= \frac{\beta_{0j}}{2} - \alpha_{0j}t - \frac{\beta_{2j}}{2}t^2.
\end{aligned}
\tag{8.23}
$$

Similarly,

$$w_{j+1}(t) = \frac{\beta_{0,j+1}}{2} - \alpha_{0,j+1}t - \frac{\beta_{2,j+1}}{2}t^2. \tag{8.24}$$

Now $x_j(t)$ is optimal for (8.1) for all t with $t_j \leq t \leq t_{j+1}$ and by construction it is also optimal for

$$\min\{-t\mu'x + \frac{1}{2}x'\Sigma x \mid A_jx = b_j\} \tag{8.25}$$

for all $t \geq 0$, where A_j and b_j correspond to the active constraints for the interval $[t_j,\ t_{j+1}]$ determined by PQPSolver. Similarly, $x_{j+1}(t)$ is optimal for

$$\min\{ -t\mu'x + \frac{1}{2}x'\Sigma x \mid A_{j+1}x = b_{j+1} \} \tag{8.26}$$

for all $t \geq 0$, where A_{j+1} and b_{j+1} correspond to the active constraints for the interval $[t_{j+1},\ t_{j+2}]$ determined by PQPSolver.

The analysis now continues according to how the end of the interval $[t_j,\ t_{j+1}]$ is determined in Step 3 of PQPSolver. If $t_{j+1} = \tilde{t}_{j+1}$, then constraint k_{l_1} is deleted from the active set for the next interval. If $t_{j+1} = \hat{t}_{j+1}$, then constraint l_2 is added to the active set for the next interval. We consider these two possibilities separately.

Case 1: $t_{j+1} = \tilde{t}_{j+1}$. In this case, a constraint which was active in the interval $[t_j,\ t_{j+1}]$ is deleted for the interval $[t_{j+1},\ t_{j+2}]$. This implies that the feasible region for (8.25) is a subset of the feasible region for (8.26). Furthermore, the optimal solution x_j for the interval $[t_j,\ t_{j+1}]$ is feasible for (8.26). Therefore, $w_j(t) \geq w_{j+1}(t)$ for all t. With (8.23) and (8.24), this implies

$$\frac{\beta_{0j}}{2} - t\alpha_{0j} - \frac{\beta_{2j}}{2}t^2 \geq \frac{\beta_{0,j+1}}{2} - t\alpha_{0,j+1} - \frac{\beta_{2,j+1}}{2}t^2, \tag{8.27}$$

for all $t \geq t_{j+1}$. When t is very large in (8.27), the quadratic terms in t will dominate. Consequently,

$$\beta_{2j} \leq \beta_{2,j+1}. \tag{8.28}$$

Using (8.28) in (8.22) gives

$$\beta_{0j} \geq \beta_{0,j+1}. \tag{8.29}$$

Finally, using $\beta_{2j} = \alpha_{1j}$, $\beta_{2,j+1} = \alpha_{1,j+1}$ and (8.28) in (8.21) gives

$$\alpha_{0j} \geq \alpha_{0,j+1}. \tag{8.30}$$

Case 2: $t_{j+1} = \hat{t}_{j+1}$. In this case, a constraint which was inactive in the interval $[t_j,\ t_{j+1}]$ becomes active in the interval $[t_{j+1},\ t_{j+2}]$. Thus, the feasible region for (8.26) is a subset of the feasible region for (8.25). In addition, the optimal solution x_{j+1} for the interval $[t_{j+1}, \ t_{j+2}]$ is feasible for (8.25). Therefore, $w_{j+1}(t) \geq w_j(t)$ for all $t \geq t_{j+1}$. With (8.23) and (8.24), this implies

$$\frac{\beta_{0,j+1}}{2} \ - \ t\alpha_{0,j+1} \ - \ \frac{\beta_{2,j+1}}{2}t^2 \geq \frac{\beta_{0j}}{2} \ - \ t\alpha_{0j} \ - \ \frac{\beta_{2j}}{2}t^2, \quad (8.31)$$

for all $t \geq t_{j+1}$. When t is very large in (8.31), the quadratic terms in t will dominate. Consequently,

$$\beta_{2j} \geq \beta_{2,j+1}. \quad (8.32)$$

Using (8.32) in (8.22) gives

$$\beta_{0j} \leq \beta_{0,j+1}. \quad (8.33)$$

Finally, using $\beta_{2j} = \alpha_{1j}$, $\beta_{2,j+1} = \alpha_{1,j+1}$ and (8.32) in (8.21) gives

$$\alpha_{0j} \leq \alpha_{0,j+1}. \quad (8.34)$$

We summarize our results in the following theorem.

Theorem 8.2 *Let PQPSolver be applied to (8.1), let $t_0 = 0$ and for $j = 0, 1, \ldots, N - 1$ let t_{j+1}, h_{0j}, h_{1j}, be the results so obtained, assume that $\tilde{t}_{j+1}, \hat{t}_{j+1}$ and t_{j+1} are uniquely determined, and let $\alpha_{0j}, \alpha_{1j}, \beta_{0j}$ and β_{2j} be as in Theorem 8.1. Then for $j = 0, 1, \ldots, N - 1$ the following hold.*

(a) If $t_{j+1} = \tilde{t}_{j+1}$, then $\beta_{2j} \leq \beta_{2,j+1}$, $\beta_{0j} \geq \beta_{0,j+1}$ and $\alpha_{0j} \geq \alpha_{0,j+1}$.

(b) If $t_{j+1} = \hat{t}_{j+1}$, then $\beta_{2j} \geq \beta_{2,j+1}$, $\beta_{0j} \leq \beta_{0,j+1}$ and $\alpha_{0j} \leq \alpha_{0,j+1}$.

Example 8.2
Illustrate Theorem 8.2 as it applies to the results of Example 7.1 summarized in Table 7.2.

In this example, $t_1 = \hat{t}_1$. Note that

$$\beta_{20} = 0.0736 > 0.008 = \beta_{21}, \ \beta_{00} = 0.0055 < 0.012 = \beta_{01} \text{ and}$$

$$\alpha_{00} = 1.0673 < 1.088 = \alpha_{01},$$

which is in agreement with Theorem 8.2(b) for $j = 0$. Additionally, $t_2 = \hat{t}_2$ in this example. Note that

$$\beta_{21} = 0.008 > 0.0 = \beta_{22}, \ \beta_{01} = 0.0012 < 0.03 = \beta_{02} \text{ and}$$

$$\alpha_{01} = 1.088 < 1.1 = \alpha_{02},$$

in agreement with Theorem 8.2(b) for $j = 1$.   ◇

Continuing the discussion prior to Theorem 8.2, we observed that $x_j(t) = h_{0j} + th_{1j}$ is optimal for (8.1) for all t with $t_j \le t \le t_{j+1}$. The parameters defining this piece of the efficient frontier are α_{0j}, α_{0j}, β_{0j} and β_{2j}, with $\alpha_{0j} = \beta_{2j}$. From (2.18), the corresponding piece of the efficient frontier is

$$\sigma_p^2 - \beta_{0j} = (\mu_p - \alpha_{0j})^2 / \alpha_{1j}, \tag{8.35}$$

for μ_p satisfying

$$\alpha_{0j} + t_j \alpha_{0j} \le \mu_p \le \alpha_{0j} + t_{j+1} \alpha_{1j}, \tag{8.36}$$

or equivalently, for σ_p^2 satisfying

$$\beta_{0j} + t_j^2 \beta_{0j} \le \sigma_p^2 \le \beta_{0j} + t_{j+1}^2 \beta_{2j}. \tag{8.37}$$

Equation (8.35), with the restrictions (8.36) and (8.37), define the piece of the efficient frontier for (8.1) in the j-th interval $[t_j, \ t_{j+1}]$. However, if we ignore the restrictions (8.36) and (8.37), (8.35) becomes the entire efficient frontier for (8.25). In this case, we call (8.35) the *extension* of the efficient frontier for the j-th interval of (8.1). Recall that the coordinates of the minimum variance portfolio are (β_0, α_0) in (σ_p^2, μ_p) space.

Theorem 8.2(a) asserts that if the end of parametric interval j is determined by a previous active constraint becoming inactive ($t_{j+1} =$

$\tilde{t}_{j+1}$) then the minimum variance portfolio for the implied efficient frontier for interval $j+1$ is greater in both coordinates, than that for interval j. Furthermore, in this case $\alpha_{2j} \leq \alpha_{2,j+1}$ ($\alpha_{1j} = \beta_{2j}$ and $\alpha_{1,j+1} = \beta_{2,j+1}$) so that the rate of increase of μ_p in interval $j+1$ is greater than in interval j.

Theorem 8.2(b) asserts that if the end of parametric interval j is determined by a previously inactive constraint becoming active ($t_{j+1} = \hat{t}_{j+1}$) then the minimum variance portfolio for the implied efficient frontier for interval $j+1$ is lesser in both coordinates, than that for interval j. Furthermore, in this case $\alpha_{2j} \geq \alpha_{2,j+1}$ ($\alpha_{1j} = \beta_{2j}$ and $\alpha_{1,j+1} = \beta_{2,j+1}$) so that the rate of increase of μ_p in interval $j+1$ is less than in interval j.

8.2   Computer Results

Figure 8.1 shows the MATLAB program to solve Example 8.1. Lines 2 to 9 set up the data for this example and do the usual checking. The point x0 is just an arbitrary feasible point for the constraints of the example and will be used as a starting point for QPSolver (called within PQPSolver). Lines 10 and 11 make the call to PQPSolver. The output quantities H0j, H1j, U0j, U1j, tvec and numint contain the quantities found in PQPSolver as follows.

$$\begin{aligned}
\text{H0j} &= [h_{00},\ h_{01}, \ldots, h_{0N-1}], \\
\text{H1j} &= [h_{10},\ h_{11}, \ldots, h_{1N-1}], \\
\text{U0j} &= [u_{00},\ u_{01}, \ldots, u_{0N-1}], \\
\text{U1j} &= [u_{10},\ u_{11}, \ldots, u_{1N-1}],
\end{aligned}$$

and
$$\text{tvec} = (t_0, t_1, \ldots, t_N).$$

and numint $= N$ (of Theorem 8.1). Note that the columns of U0j and U1j give the columns of the full multiplier vectors and the notation $u_j(t) = u_{0j} + tu_{1j}$ for them is as used in Chapter 7.1.

```
1   %Example8p1.m
2   A = [ −1 0 0; 0 −1 0; 0 0 −1; 1 1 1 ];
3   b = [ 0 0 0 1 ]';
4   mu = [ 1.05 1.08 1.1 ]';
5   Sigma = [1 0 0; 0 2 0; 0 0 3]/100.;
6   n = 3;   m = 3;   q = 1;
7   x0 = [0.2 0.2 0.6 ]';
8   checkdata(Sigma,1.e−6);
9   checkconstraints(A,b,x0,m,q,n);
10  [H0j,H1j,U0j,U1j,tvec,numint] = ...
11       PQPSolver(A,b,m,n,q,x0,Sigma,mu);
12  msg = 'Efficient Frontier Completely Determined'
13  tvec
14  H0j
15  H1j
16  U0j
17  U1j
18  Alpha0 = []; Alpha1 = []; Beta0 = []; Beta2 = [];
19  for j = 1:numint
20      Alpha0(j) = mu' * H0j(1:n,j);
21      Alpha1(j) = mu' * H1j(1:n,j);
22      Beta0(j)  = H0j(1:n,j)' * Sigma * H0j(1:n,j);
23      Beta2(j)  = H1j(1:n,j)' * Sigma * H1j(1:n,j);
24  end
25  Alpha0
26  Alpha1
27  Beta0
28  Beta2
29  %  Expected return and variance at start of each interval
30  mu = [];   sig2 = [];
31  for j = 1:numint
32      mu(j)   =  Alpha0(j) + tvec(j)*Alpha1(j);
33      sig2(j) =  Beta0(j) + tvec(j)*tvec(j)*Beta2(j);
34  end
35  mu
36  sig2
```

Figure 8.1   Example 8.1: Example8p1.m.

Lines 13 to 17 print out these quantities.

Having computed h_{0j} and h_{1j} for all j, the program now computes (lines 18 to 24) the constants which define the efficient frontier in each interval. These are

$$
\begin{aligned}
\text{Alpha0} &= \left[\mu' h_{00},\ \mu' h_{01},\ \ldots,\ \mu' h_{0N-1}\right], \\
\text{Alpha1} &= \left[\mu' h_{10},\ \mu h_{11},\ \ldots,\ \mu' h_{1N-1}\right], \\
\text{Beta0} &= \left[h_{00}\Sigma h_{00},\ h_{01}\Sigma h_{01},\ \ldots,\ h'_{0,N-1}\Sigma h_{0,N-1}\right], \\
\text{Beta1} &= \left[h_{10}\Sigma h_{10},\ h_{11}\Sigma h_{11},\ \ldots,\ h'_{1,N-1}\Sigma h_{1,N-1}\right].
\end{aligned}
$$

These vectors are printed in lines 25 to 28.

Finally, lines 29 through 36 compute the mean and variance at the end of each interval and then print them.

Figure 8.2 shows the file PQPSolver.m. The input arguments are A, b, m, n, q, x0, Sigma and mu which all have their usual meanings. The output arguments H0j, H1j, U0j, U1j, tvec and numint have meanings as described for Example 8.1 (see Figure 8.1). Line 5 initializes c to zero (i.e., $t = 0$). Lines 7 and 8 invoke QPSolver to solve the QP for $t = 0$. Line 8 initializes most of the output arrays to null. Line 10 starts the flag loop. The variable flag has been initialized to zero in line 4. It will be set to 1 when the end of the parametric intervals is found. When QPsolver was called in lines 7 and 8, it returned the index set Kj (K_j) for the optimal solution when t = 0. In line 14, active is the number of active constraints (n_j). Lines 15 to 17 construct the quantities Aj (A_j), Hj (Hj) and bj (b_j) used in Step 1 of PQPSolver. Lines 19 and 20 form the right-hand sides of the linear equations to be solved in Step 1. Line 21 solves these equations and lines 22 to 23 extract h0j (h_{0j}), h1j (h_{1j}), g0j (g_{0j}) and g1j (g_{1j}) from them. Line 24 initializes u0j (u_{0j}) and u1j (u_{1j}) to zero. Lines 26 and 27 fill in the nonzero components of these two vectors using g0j and g1j, respectively and the active set Kj (K_j). Lines 29 and 30 then augment H0j, H1j, U0j and U1j with the new last columns h0j, h1j, u0j and u1j, respectively.

Step 2 of PQPSolver begins at line 32. Lines 32 to 40 compute $\tilde{t}_{j+1}$ as follows. Line 33 computes set, the set of indices of components of g1j which are $\leq tolstp$. Note that the indices considered are (active–q). Because the equality constraints are always placed as the bottom rows of Aj, this means the multipliers for the equality constraints will never be considered. This is as it should be since the calculation of $\tilde{t}_{j+1}$ in

Step 2 of PQPSolver has the requirement that only active inequality constraints $(1 \leq k_i \leq m)$ be considered. If set is nonempty, test is set to the vector of candidate ratios (line 35), then tjtilde is set to the minimum of these (line 36) and ell1 is the index for which the minimum occurs (line 37). If there are no candidates, then tjtilde is set to ∞.

Lines 42 to 53 compute $\hat{t}_{j+1}$ as follows. The set inactive (line 43) consists of the indices of those constraints which are inactive at the current point and the vector sumh (line 44) contains the inner product of the gradients of each such constraint with the vector h1j. In line 45, ind is computed as the set of indices of those inner products which are greater than the tolerance tolstp (a number slightly greater than zero, defined in line 4). Lines 46 to 53 are executed if the set ind is nonnull. In this case, the vector tjhattest (line 47) is the set of all candidate ratios for $\hat{t}_{j+1}$. In line 49, tjhat is computed as the smallest of these and ell2 (line 50) is the corresponding index of the constraint for which the minimum occurs. If the set ind is empty, tjhat is set to ∞ in line 52.

Lines 55 to 75 update the index set Kj as follows. First, line 56 computes tj1 $(= t_{j+1})$ as the minimum of tjtilde $(= \tilde{t}_{j+1})$ and tjhat $(= \hat{t}_{j+1})$. This value is then stored in tvec in line 57. Lines 58 to 62 check to see if tj1 $= \infty$ and if so numint is set to j, flag is set to 1 (indicating that all intervals have been found) and control returns to the calling program.

If tj1 = tjtilde $(t_{j+1} = \tilde{t}_{j+1})$, then Kj is modified by removing element ell1 and active is reduced by one (lines 65 and 66). Thus the constraint with index KJ(ell1) has been removed from the active set.

If tj1 = tjhat $(t_{j+1} = \hat{t}_{j+1})$, then ell2 is added to the start of Kj (lines 70 and 71). Thus constraint with index ell2 has been added to the active set. In addition, lines 72 to 74 check if the number of active constraints is n. If so, then the current point is an extreme point and this information is printed.

Finally, line 76 makes the start of the next interval the end of the present interval, sets numint to be the number of intervals recorded by the interval counter j and then increases j by 1 for the next parametric interval.

```
1  %PQPSolver.m
2  function [H0j,H1j,U0j,U1j,tvec,numint] = ...
3      PQPSolver(A,b,m,n,q,x0,Sigma,mu)
4  tolstp = 1.e-8;   flag = 0;   j = 1;
5  c = 0 .* mu;
6  BigC =  Sigma;
7  [minvalue,xj,Kj,uj] = QPSolver(A,b,BigC,c,n,m,q,x0);
8  H0j = []; H1j = []; U0j = []; U1j = []; tj = 0.;KJ = [];
9  tvec(1) = 0;
10 while flag == 0
11
12 %Construct Aj from active set Kj
13     Aj = []; bj = [];   rhs1 = []; rhs2 = [];
14     active = size(Kj',1);
15     Aj = A(Kj,:);
16     Hj = vertcat([Sigma Aj'],[Aj zeros(active,active)]);
17     bj = b(Kj);
18
19     rhs1 = vertcat(zeros(n,1), bj);
20     rhs2 = vertcat(mu, zeros(active,1));
21     y1 = Hj^-1 * rhs1;        y2 = Hj^-1 * rhs2;
22     h0j = y1(1:n);           h1j = y2(1:n);
23     g0j = y1(n+1:n+active); g1j = y2(n+1:n+active);
24     u0j = zeros(m+q,1);      u1j = zeros(m+q,1);
25
26     u0j(Kj(1:active)) = g0j(1:active);
27     u1j(Kj(1:active)) = g1j(1:active);
28
29     H0j = horzcat(H0j,h0j);   H1j = horzcat(H1j,h1j);
30     U0j = horzcat(U0j,u0j);   U1j = horzcat(U1j,u1j);
31
32 % limit on t for multipliers
33     set = find(g1j(1:active-q)<-tolstp);
34     if ¬isempty(set);
35       test = - g0j(set)/g1j(set);
36       tjtilde = min(test);
37       ell1 = set(find(test==tjtilde,1));
38     else
39        tjtilde = Inf;
40     end
41
42 % limit on t for constraints
43     inactive = setdiff(1:m,Kj);
44     sumh = A(inactive,:)*h1j;
45     ind = find(sumh>tolstp);
```

```
46      if ¬isempty(ind)
47          tjhattest = (b(inactive(ind)) − ...
48                      A(inactive(ind),:)*h0j)  ./sumh(ind);
49          tjhat = min(tjhattest);
50          ell2 = inactive(ind(find(tjhattest==tjhat,1)));
51      else
52          tjhat = Inf;
53      end;
54
55  %update index set Kj
56      tj1 = min(tjtilde,tjhat);
57      tvec(j+1) = tj1;
58      if tj1 == Inf
59          numint = j;
60          flag = 1;
61          return
62      end
63
64      if tj1 == tjtilde
65          Kj(ell1) = [];
66          active = active − 1;
67      end
68
69      if tj1 == tjhat
70          active = active + 1;
71          Kj = [ell2 Kj];
72          if active == n
73              msg = 'Extreme Point'
74          end
75      end
76      tj = tj1;   numint = j;   j = j + 1;
77      Kj
78      cornor = h0j + tj1 *h1j
79  end
```

Figure 8.2   PQPSolver.m.

8.3 Exercises

8.1 Use the data from Example 4.1 with PQPSolver to find the entire efficient frontier for a problem with nonnegativity constraints on all assets. Summarize your results in a table similar to Table 7.2. Also make a table showing the corner portfolios. Use PQPSolver to perform the calculations.

8.2 Repeat Exercise 8.1, but with a risk free asset having $r = 1.02$. Note that the resultant covariance for this problem will have a zero row and column and so will not be positive definite. However, because of the problem's special structure, PQPSolver will indeed solve it. See the discussion in Section 9.1.

8.3 Consider a problem with n risky assets, asset $(n+1)$ being risk free and with nonnegativity constraints on all $n+1$ assets; i.e.,

$$\min\{-t(\mu'x+rx_{n+1})+\frac{1}{2}x'\Sigma x \mid l'x+x_{n+1} = 1,\ x \geq 0,\ x_{n+1} \geq 0\}.$$

(a) What is the optimal solution when $t = 0$? Which constraints are active at it?

(b) Can PQPSolver be applied to solve it? Why or why not?

(c) Let h_1 be the optimal solution for

$$\min\{-(\mu-rl)'h_1 + \frac{1}{2}h_1'\Sigma h_1 \mid h_1 \geq 0\}.$$

Show that $[(0', 1) + t(h_1', 1 - l'h_1)]'$ is optimal for the given problem for all t sufficiently small but positive.

Chapter 9

Sharpe Ratios under Constraints, and Kinks

Practical portfolio optimization requires constraints. These may be lower bound constraints to control short selling and upper bound constraints to ensure that reoptimized portfolios do not move too far from the original portfolio. Practitioners may use linear inequality constraints to model various aspects of the theory they are implementing. In any case, linear inequality constraints are essential to realistic portfolio optimization.

In this chapter, we will generalize many of the results of Sections 4.2 and 4.3 (Sharpe ratios and implied risk free returns) to problems having general linear inequality constraints. We assume throughout this chapter that the covariance matrix Σ is positive definite.

9.1 Sharpe Ratios under Constraints

In this section, we will relate portfolios obtained from constrained versions of the defining portfolios (2.1), (2.2) and (2.3). These are:

$$\min\{x'\Sigma x \mid \mu'x = \mu_p, \quad Ax \leq b\}, \tag{9.1}$$

$$\max\{\mu'x \mid x'\Sigma x = \sigma_p^2, \ Ax \le b\}, \tag{9.2}$$

and

$$\min\{-t\mu'x + \frac{1}{2}x'\Sigma x \mid Ax \le b\}. \tag{9.3}$$

In each of these problems, A is an (m, n) matrix and b is an m-vector of corresponding right-hand sides. These constraints could represent upper and lower bounds, sector constraints or any other constraints the practitioner chooses to incorporate into the model. We implicitly allow these constraints to include equality constraints. This allows for the budget constraint. From Theorem 5.2, the optimality conditions for an equality constraint require no sign restriction for the associated multiplier, the equality constraint is always active, and there is no complementary slackness condition for it. The optimality conditions for (9.1) are

$$\mu'x = \mu_p, \ Ax \le b,$$
$$-2\Sigma x = v\mu + A'u_1, \ u_1 \ge 0, \tag{9.4}$$
$$u_1'(Ax - b) = 0.$$

Here, v is the (scalar) multiplier for the constraint $\mu'x = \mu_p$ and u_1 is the vector of multipliers for the constraints $Ax \le b$.

The optimality conditions for (9.2) (see Exercise 2.4) are

$$x'\Sigma x = \sigma_p^2, \ Ax \le b,$$
$$\mu = 2w\Sigma x + A'u_2, \ u_2 \ge 0, \tag{9.5}$$
$$u_2'(Ax - b) = 0.$$

Here, w is the (scalar) multiplier for the constraint $x'\Sigma x = \sigma_p^2$ and u_2 is the vector of multipliers for the constraints $Ax \le b$.

Finally, the optimality conditions for (9.3) are

$$Ax \le b,$$
$$t\mu - \Sigma x = A'u_3, \ u_3 \ge 0, \tag{9.6}$$
$$u_3'(Ax - b) = 0.$$

Here, u_3 is the vector of multipliers for the constraints $Ax \le b$.

Now let t_1 be any positive number and let $x_1 \equiv x(t_1)$ be optimal for (9.3) when $t = t_1$. The dual feasibility part of the optimality conditions for (9.1) can be rearranged as

$$(-v/2)\mu - \Sigma x = A'(u_1/2).$$

Defining $v = -2t_1$, $u_1 = 2u_3$, $x = x_1$ and $\mu_p = \mu' x_1$ shows that x_1 satisfies all of the optimality conditions for (9.1). Therefore x_1 is optimal for (9.1) with $\mu_p = \mu' x_1$.

Again, let t_1 be any positive number and let $x_1 \equiv x(t_1)$ be optimal for (9.3) when $t = t_1$. Rearranging the dual feasibility part of the optimality conditions for (9.3) gives

$$\mu = \frac{1}{t}\Sigma x_1 + A'(\frac{1}{t}u_3).$$

By defining

$$w = \frac{1}{2t}, \quad u_2 = \frac{1}{t}u_3,$$

it follows that x_1 satisfies all of the optimality conditions for (9.2). Therefore x_1 is optimal for (9.2) with $\sigma_p^2 = x_1'\Sigma x_1$.

We summarize our results so far in the following theorem.

Theorem 9.1 *For any $t_1 > 0$, let $x_1 \equiv x(t_1)$ be an optimal solution for (9.3). Then (a) x_1 is also optimal for (9.1) with $\mu_p = \mu' x_1$, (b) x_1 is also optimal for (9.2) with $\sigma_p^2 = x_1'\Sigma x_1$.*

In Section 4.2, we studied the problem of maximizing the Sharpe ratio subject only to the budget constraint. The problem analyzed was (4.6) and we repeat it here for convenience:

$$\max \left\{ \frac{\mu' x - r_m}{(x'\Sigma x)^{\frac{1}{2}}} \mid l'x = 1 \right\}.$$

In this section, we wish to maximize the Sharpe ratio subject to general linear inequality constraints. That is, we wish to solve

$$\max \left\{ \frac{\mu' x - r_m}{(x'\Sigma x)^{\frac{1}{2}}} \mid Ax \le b \right\}. \tag{9.7}$$

In equation (4.7), it was shown that the gradient of the objective function for (9.7) is

$$\frac{(x'\Sigma x)^{\frac{1}{2}}\mu \; - \; (\mu'x \; - \; r_m)(x'\Sigma x)^{-\frac{1}{2}}\Sigma x}{x'\Sigma x}. \tag{9.8}$$

After rearrangement, the optimality conditions (see Exercise 5.11) for (9.7) are

$$Ax \le b, \tag{9.9}$$

$$\left[\frac{x'\Sigma x}{\mu'x \; - \; r_m}\right]\mu \; - \; \Sigma x \; = \; A'u_4\left[\frac{(x'\Sigma x)^{3/2}}{\mu'x \; - \; r_m}\right], \quad u_4 \ge 0, \tag{9.10}$$

and

$$u_4'(Ax - b) = 0. \tag{9.11}$$

Now let t_1 be any positive number and let $x_1 \equiv x(t_1)$ be the optimal solution of (9.3). Then x_1 satisfies the optimality conditions (9.6) with $x = x_1$ and $t = t_1$. These are

$$\begin{aligned} Ax_1 &\le b, \\ t_1\mu \; - \; \Sigma x_1 &= A'u_3, \quad u_3 \ge 0, \\ u_3'(Ax_1 - b) &= 0. \end{aligned} \tag{9.12}$$

In (9.9), (9.10) and (9.11), define $x = x_1$,

$$\left[\frac{x_1'\Sigma x_1}{\mu'x_1 \; - \; r_m}\right] = t_1 \text{ and } u_4 = \left[\frac{\mu'x_1 \; - \; r_m}{(x_1'\Sigma x_1)^{3/2}}\right]u_3. \tag{9.13}$$

Note that

$$\left[\frac{\mu'x_1 \; - \; r_m}{(x_1'\Sigma x_1)^{3/2}}\right]$$

is just a nonnegative scalar so that $u_3 \ge 0$ implies $u_4 \ge 0$. Furthermore, $u_3'(Ax_1 - b) = 0$ implies $u_4'(Ax_1 - b) = 0$.

Thus x_1 satisfies all of the optimality conditions for (9.7) and is therefore an optimal solution for it. In verifying that x_1 is optimal for (9.7), we defined

$$\left[\frac{x_1'\Sigma x_1}{\mu'x_1 \; - \; r_m}\right] = t_1.$$

Both t_1 and x_1 are fixed, but r_m is not. Therefore we can define

$$r_m = \mu' x_1 - \frac{x_1' \Sigma x_1}{t_1}, \tag{9.14}$$

and the conclusion is that any point x_1 on the efficient frontier for (9.3) (specified by t_1) gives the maximum Sharpe ratio for r_m specified by (9.14). Note that if we wish to find the maximum Sharpe ratio for a specified r_m, we would have to find the corresponding t and that is a different matter. We summarize our results in the following theorem.

Theorem 9.2 *Let t_1 be a positive number and let $x_1 \equiv x(t_1)$ be the optimal solution of (9.3) for $t = t_1$. Then x_1 gives the maximum Sharpe ratio for (9.7) for r_m given by (9.14).*

In Chapter 3, we discussed the Capital Market Line for a Mean-Variance problem having only a budget constraint. In (3.10), we required the assumption that $r < \alpha_0$ where r is the risk free return and α_0 is the expected return on the minimum variance portfolio ($\alpha_0 = \mu' h_0$). Our next result shows that the analogous result for (9.7) with r_m being given by (9.14) is always satisfied. In Theorem 9.2 and the development prior to its statement, we used t_1 to be an arbitrary positive number. In order to avoid confusion with t_1 denoting the end of a parametric interval, we shall use t^* in place of t_1 and x^* in place of $x_1 \equiv x(t_1)$.

Theorem 9.3 *Let $t_0, t_1, \ldots, N$ be as in the statement of Theorem 7.1. Let $t^* > 0$ and let j be such that $t^* \in [t_j, t_{j+1}]$. Let $\mu_{pj}, \sigma_{pj}^2, \alpha_{0j}, \alpha_{1j}, \beta_{0j}$ and β_{2j} be as in the statement of Theorem 7.1. Let r_m be obtained from (9.14) with x_1 and t_1 replaced by x^* and t^*, respectively. Then $r_m \leq \alpha_{0j}$.*

Proof: From Theorem 7.1(b), $\mu' x* = \alpha_{0j} + t^* \alpha_{1j}$ and $(x^*)' \Sigma x^* = \beta_{0j} + \beta_{2j}(t^*)^2$. Furthermore, from (7.50) we have $\alpha_{1j} = \beta_{2j}$. Therefore

$$r_m = \alpha_{0j} + t^* \alpha_{1j} - \left[\frac{\beta_{0j} + \beta_{2j}(t^*)^2}{t^*} \right]$$

$$= \alpha_{0j} + t^* \alpha_{1j} - \frac{\beta_{0j}}{t^*} - t^* \beta_{2j}$$

$$= \alpha_{0j} - \frac{\beta_{0j}}{t^*}. \tag{9.15}$$

Since $\beta_{0j} \geq 0$ and $t^* > 0$ it follows from (9.15) that $r_m \leq \alpha_{0j}$, as required. $\qquad\square$

Note that in the proof of Theorem 9.5, if $\beta_{2j} > 0$ then $r_m < \alpha_{0j}$

We next consider the problem of maximizing the implied risk free return, r_m, when the Sharpe ratio, R, is given. This was studied in Section 4.2 and its generalization to linear inequality constraints is

$$\max\{\mu'x - R(x'\Sigma x)^{\frac{1}{2}} \mid Ax \leq b\}. \tag{9.16}$$

The gradient of the objective function for this problem is

$$\mu - \left[\frac{R}{(x'\Sigma x)^{\frac{1}{2}}}\right]\Sigma x.$$

After rearrangement, the optimality conditions (see Exercise 5.11) for (9.16) are

$$Ax \leq b, \tag{9.17}$$

$$\left[\frac{(x'\Sigma x)^{\frac{1}{2}}}{R}\right]\mu - \Sigma x = A'u_5\left[\frac{(x'\Sigma x)^{\frac{1}{2}}}{R}\right], \quad u_5 \geq 0, \tag{9.18}$$

$$u_5'(Ax - b) = 0. \tag{9.19}$$

Now let t_1 be any positive number and let $x_1 \equiv x(t_1)$ be the optimal solution of (9.3). Then x_1 satisfies the optimality conditions (9.6) with $x = x_1$ and $t = t_1$. These are

$$Ax_1 \leq b,$$
$$t_1\mu - \Sigma x_1 = A'u_3, \quad u_3 \geq 0, \tag{9.20}$$
$$u_3'(Ax_1 - b) = 0.$$

In (9.17), (9.18) and (9.19), define $x = x_1$,

$$\left[\frac{(x_1'\Sigma x_1)^{\frac{1}{2}}}{R}\right] = t_1 \text{ and } u_5 = \left[\frac{R}{(x_1'\Sigma x_1)^{\frac{1}{2}}}\right] u_3. \tag{9.21}$$

Note that

$$\left[\frac{R}{(x_1'\Sigma x_1)^{\frac{1}{2}}}\right] \tag{9.22}$$

is just a nonnegative scalar so that $u_3 \geq 0$ implies $u_5 \geq 0$. Furthermore, $u_3'(Ax_1 - b) = 0$ implies $u_5'(Ax_1 - b) = 0$.

Thus x_1 satisfies all of the optimality conditions for (9.16) and is therefore an optimal solution for it. In verifying that x_1 is optimal for (9.16), we defined

$$\left[\frac{(x_1'\Sigma x_1)^{\frac{1}{2}}}{R}\right] = t_1.$$

Both t_1 and x_1 are fixed, but R is not. Therefore we can define

$$R = \left[\frac{(x_1'\Sigma x_1)^{\frac{1}{2}}}{t_1}\right] \tag{9.23}$$

and the conclusion is that any point x_1 on the efficient frontier for (9.3) (specified by t_1) gives the implied risk free return for the Sharpe ratio specified by (9.23). Note that if we wish to find the implied risk free return for a specified R, we would have to find the corresponding t and that is a different matter. We summarize our results in the following theorem.

Theorem 9.4 *Let t_1 be a positive number and let $x_1 \equiv x(t_1)$ be the optimal solution of (9.3) for $t = t_1$. Then x_1 gives the optimal solution for (9.16) for R given by (9.23).*

In Chapter 3, we discussed the Capital Market Line for a Mean-Variance problem having only a budget constraint. In (4.21), we required the assumption that $R \geq \beta_0^{\frac{1}{2}}$ where R is the slope of the CML and β_0 is the variance of the minimum variance portfolio ($\beta_0 = h_0' \Sigma h_0$). Our next result shows that the analogous result for (9.7) with R being given by (9.23) is always satisfied. In Theorem 9.4 and the development prior to its statement, we used t_1 to be an arbitrary positive number. In order to avoid confusion with t_1 denoting the end of a parametric interval, we shall use t^* in place of t_1 and x^* in place of $x_1 \equiv x(t_1)$.

Theorem 9.5 *Let $t_0, t_1, \ldots, N$ be as in the statement of Theorem 7.1. Let $t^* > 0$ and let j be such that $t^* \in [t_j, t_{j+1}]$. Let μ_{pj}, σ_{pj}^2, α_{0j}, α_{1j}, β_{0j} and β_{2j} be as in the statement of Theorem 7.1. Let R be obtained from (9.23) with x_1 and t_1 replaced by x^* and t^*, respectively. Then $R \geq \beta_{0j}^{\frac{1}{2}}$.*

Proof: From Theorem 7.1(b), $(x^*)' \Sigma x^* = \beta_{0j} + \beta_{2j}(t^*)^2$. Therefore

$$R = \frac{((x^*)' \Sigma x^*)^{\frac{1}{2}}}{t^*} = \frac{\beta_{0j} + (t^*)^2}{t^*}.$$

Cross multiplying and squaring gives

$$(t^*)^2 R^2 = \beta_{0j} + (t^*)^2 \beta_{2j},$$

or,

$$R^2 = \beta_{2j} + \frac{\beta_{0j}}{(t^*)^2}.$$

Since $\beta_{0j} \geq 0$ and $(t^*)^2 > 0$, the result follows.   □

Note that in the proof of Theorem 9.5, if $\beta_{0j} > 0$, then $R > \beta_{2j}$.

Example 9.1
Apply Theorems 9.1, 9.2 and 9.4 to the problem of Example 7.1 with $t_1 = 1$. Use the optimal solution for it as summarized in Table 7.2.

When $t_1 = 1$, Table 7.2 gives the optimal solution as

$$x_1 = \begin{bmatrix} 0.0 \\ 0.6 \\ 0.4 \end{bmatrix} + 1 \times \begin{bmatrix} 0.0 \\ -0.4 \\ 0.4 \end{bmatrix} = \begin{bmatrix} 0.0 \\ 0.2 \\ 0.8 \end{bmatrix}.$$

Furthermore, from Table 7.2 the corresponding expected return is

$$\mu_p = \mu' x_1 = 1.088 + 1 \times 0.008 = 1.096$$

and the variance is

$$\sigma_p^2 = x_1' \Sigma x_1 = 0.012 + 1^2 \times 0.008 = 0.02.$$

From Theorem 9.1, x_1 is optimal for (9.1) with $\mu_p = 1.096$ and is also optimal for (9.2) with $\sigma_p^2 = 0.02$.

To apply Theorem 9.2, we need to calculate r_m according to (9.14). Doing so gives

$$r_m = \mu' x_1 - \frac{x_1' \Sigma x_1}{t_1} = 1.096 - 0.02 = 1.076.$$

Theorem 9.2 now asserts that x_1 gives the maximum Sharpe ratio with $r_m = 1.076$. That is, x_1 is the optimal solution for (9.7) with $r_m = 1.076$.

To apply Theorem 9.4, we need to calculate R according to (9.23). Doing so gives

$$R = \left[\frac{(x_1' \Sigma x_1)^{\frac{1}{2}}}{t_1} \right] = (0.02)^{\frac{1}{2}} = 0.1414.$$

Theorem 9.4 now asserts that x_1 gives the optimal solution for (9.16) when $R = 0.1414$.

9.2 Kinks and Sharpe Ratios

We continue using the model problem (7.2) of Chapter 7 and repeat it here for convenience:

$$\begin{aligned} \text{minimize} : & \quad -t\mu' x + \tfrac{1}{2} x' \Sigma x \\ \text{subject to} : & \quad a_i' x \leq b_i, \ i = 1, 2, \ldots, m, \\ & \quad a_i' x = b_i, \ i = m + 1, m + 2, \ldots, m + q. \end{aligned} \quad (9.24)$$

In Chapter 7, Theorem 7.1 proved the efficient frontier for (9.24) is piece wise parabolic and differentiable for all values of t with the possible exception of those t_j for which $\beta_{2j} = 0$. These last points are called **kink points**. In this chapter, we will examine kink points in detail. We will show how such points arise, show that the efficient frontier becomes nondifferentiable at each kink point, and show the effect they have on Sharpe ratios and the implied risk free return.

We first illustrate many of the most important ideas through the following simple example.

Example 9.2

Consider the 3 asset portfolio optimization problem with $\mu = (1.05, 1.08, 1.1)'$, $\Sigma = \text{diag}(1, 2, 3)/100$ and constraints

$$
\begin{array}{rcccl}
x_1 & & & \leq & 0.8, \quad (1) \\
 & x_2 & & \leq & 0.25, \quad (2) \\
 & & x_3 & \leq & 0.8, \quad (3) \\
 & x_2 + & x_3 & \leq & 0.85, \quad (4) \\
x_1 + & x_2 + & x_3 & = & 1. \quad (5)
\end{array}
$$

We can generate the entire efficient frontier for this problem using PQP-Solver (see Chapter 8.1 and the MATLAB program presented in Figure 8.2). The MATLAB program to set up the data and call PQPSolver is shown in Figure 9.4. The output from that program is summarized in Table 9.1. This table also shows the coefficients of the efficient frontier α_{0j}, α_{1j}, β_{0j} and β_{2j} for $j = 0, 1, \ldots, 3$. This information is used to graph the efficient frontier. Figure 9.1 shows the entire efficient frontier in Mean-Variance space and Figure 9.2 shows it in Mean-Standard Deviation space. At the indicated "kink" point, the efficient frontier is not differentiable. From Table 9.1, we see that this occurs for t in the interval $[0.33, 0.65]$, where $t_1 = 0.33$ and $t_2 = 0.65$. In this interval, $\beta_{21} = 0$ and since $\beta_{21} = h'_{11}\Sigma h_{11}$ and Σ is positive definite, it follows that $h_{11} = 0$. This is in agreement with Table 9.1. Thus the efficient portfolios for this interval are

$$
x(t) = \begin{bmatrix} 0.15 \\ 0.25 \\ 0.60 \end{bmatrix} + t \begin{bmatrix} 0.0 \\ 0.0 \\ 0.0 \end{bmatrix} = \begin{bmatrix} 0.15 \\ 0.25 \\ 0.60 \end{bmatrix}. \tag{9.25}
$$

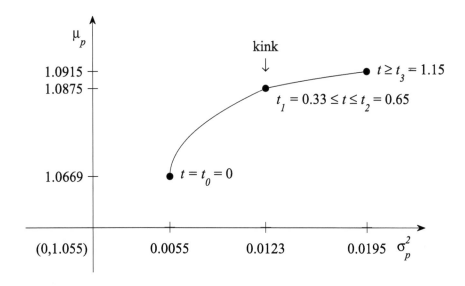

Figure 9.1   Example 9.2: Kink in Mean-Variance Space.

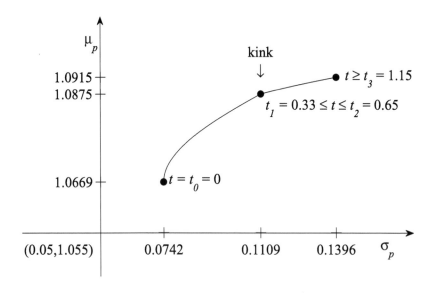

Figure 9.2   Example 9.2: Kink in Mean-Standard Deviation Space.

TABLE 9.1   **Piecewise Linear, Optimal Solution for Example 9.2 and Associated Efficient Frontier Parameters**

j	0	1	2	3
t_j	0	0.33	0.65	1.15
K_j	$\{5,2\}$	$\{5,2,4\}$	$\{5,4\}$	$\{5,4,3\}$
	h_{00}	h_{01}	h_{02}	h_{03}
	0.5625	0.1500	0.1500	0.1500
h_{0j}	0.2500	0.2500	0.5100	0.0500
	0.1875	0.6000	0.3400	0.8000
	h_{10}	h_{11}	h_{12}	h_{13}
	-1.2500	0.0	0.0	0.0
h_{1j}	0.0	0.0	-0.400	0.0
	1.2500	0.0	0.400	0.0
	u_{00}	u_{01}	u_{02}	u_{03}
	0.0	0.0	0.0	0.0
	0.0006	0.0130	0.0	0.0
u_{0j}	0.0	0.0	0.0	-0.0230
	0.0	-0.0165	-0.0087	0.0005
	-0.0056	-0.0015	-0.0015	-0.0015
	u_{10}	u_{11}	u_{12}	u_{13}
	0.0	0.0	0.0	0.0
	0.0175	-0.02	0.0	0.0
u_{1j}	0.0	0.0	0.0	0.02
	0.0	0.05	0.038	0.03
	1.0625	1.05	1.05	1.05
α_{0j}	1.0669	1.0875	1.0823	1.0915
α_{1j}	0.0625	0.0	0.0080	0.0
β_{0j}	0.0055	0.0123	0.0089	0.0195
β_{2j}	0.0625	0.0	0.0080	0.0

Furthermore, from Table 9.1 the corresponding expected return and variance in this interval are

$$\mu_p = 1.0875 + 0t = 1.0875 \text{ and } \sigma_p^2 = 0.0123 + 0t^2 = 0.0123.$$

(9.26)

Equations (9.25) and (9.26) show that $x(t), \mu_p$ and σ_p^2 remain constant over the interval for t of $[0.33, 0.65]$. Thus the efficient frontier has been reduced to a single point for t in this interval. Observe there are three constraints active at $x(t)$, namely constraints (2), (4), and (5). By Definition 1.4 the corresponding efficient portfolio $x_1 = (0.15, 0.25, 0.60)'$ is an extreme point.

For Example 9.2, the corner portfolios (see (7.57)) are shown in Table 9.2.

TABLE 9.2   **Corner Portfolios for Example 9.2**

j	0	1	2	3
t_j	0	0.33	0.65	1.15
K_j	$\{5, 2\}$	$\{5, 2, 4\}$	$\{5, 4\}$	$\{5, 4, 3\}$
$\mu_p(t_j)$	1.0669	1.0875	1.0875	1.0915
$\sigma_p^2(t_j)$	0.0055	0.0123	0.0123	0.0195
	$\hat{x}_0$	$\hat{x}_1$	$\hat{x}_2$	$\hat{x}_3$
	0.5625	0.1500	0.1500	0.1500
$\hat{x}_j$	0.2500	0.2500	0.2500	0.0500
	0.1875	0.6000	0.6000	0.8000

For any t in the interval $0.33 \le t \le 0.65$, the optimal solution for this problem is
$$x_1 = (0.15, 0.25, 0.60)'.$$

Theorem 9.2 asserts that x_1 gives the maximum Sharpe ratio with

$$
\begin{aligned}
r_m &= \mu' x_1 - \left[\frac{x_1' \Sigma x_1}{t} \right] \\
&= 1.0875 - \frac{0.0123}{t},
\end{aligned}
\tag{9.27}
$$

for all t in the interval $[0.33, 0.65]$. This means that when the maximum Sharpe ratio occurs at a kink, there is an interval of implied risk free returns corresponding to it. For the example here, the smallest of these risk free returns corresponds to $t_1 = 0.33$ and is

$$r_{m1} = 1.0875 - \frac{0.0123}{0.33} = 1.0502, \tag{9.28}$$

and the largest corresponds to $t_2 = 0.65$ and is

$$r_{m2} = 1.0875 - \frac{0.0123}{0.65} = 1.0686. \tag{9.29}$$

This is illustrated in Figure 9.3. Note that the line joining the kink point $r_{m1} = 1.0502$ is the left-hand tangent to the efficient frontier at t_1 and the line joining $r_{m2} = 1.0686$ is the right-hand tangent to the efficient frontier at t_2. Note that all implied risk free returns in the interval $[r_{m1}, r_{m2}]$ are below α_{01}, in agreement with Theorem 9.5.

Theorem 9.4 can also be applied to this example. Theorem 9.4 states that for all t in the interval $[0.33, 0.65]$

$$R = \left[\frac{(x_1' \Sigma x_1)^{\frac{1}{2}}}{t} \right]$$

$$= \left[\frac{0.1109}{t} \right], \tag{9.30}$$

is the Sharpe ratio for which the risk free return

$$r_m = \mu' x_1 - R(x_1' \Sigma x_1)^{\frac{1}{2}}$$
$$= 1.0875 - 0.1109R \tag{9.31}$$

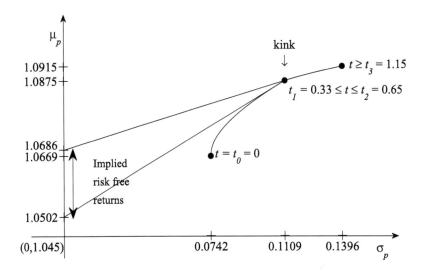

Figure 9.3.   Example 9.2: Risk Free Returns Implied By Kinked Market Portfolio

is greatest. Substituting (9.30) into (9.31) gives

$$
\begin{aligned}
r_m &= 1.0875 - \left[\frac{0.1109}{t}\right] \times 0.1109 \\
&= 1.0875 - \left[\frac{0.0123}{t}\right],
\end{aligned}
\tag{9.32}
$$

in agreement with (9.27).

Example 9.3

This example is based on Example 9.2 and has a risk free asset, numbered 4, and includes the constraint $x_4 \geq 0$. The risk free return is $r = 1.01$.

     The covariance matrix for the three risky assets is positive definite. However, including a risk free asset gives a $(4, 4)$ covariance matrix in which the last row and column are all zeros and is thus positive semidefinite but not positive definite. Nonetheless, both QPSolver and

PQPSolver can both solve problems of this form provided the budget constraint is included in the problem being solved. The key issue is that the matrices H_j are nonsingular. See Exercise 9.2.

The MATLAB code to find the entire efficient frontier using PQP-Solver is shown in Figure 9.5. The results of running it are summarized in Table 9.3. In addition, the corner portfolios, their expected returns and variances are shown in Table 9.4.

TABLE 9.3   **Piecewise Linear, Optimal Solution for Example 9.3 and Associated Efficient Frontier Parameters**

j	0	1	2	3	4	5
t_j	0	0.0714	0.1071	0.33	0.6500	1.15
K_j	$\{6\}$	$\{2,6\}$	$\{5,2,6\}$	$\{4,5,2,6\}$	$\{4,5,6\}$	$\{3,4,5,6\}$
	h_{00}	h_{01}	h_{02}	h_{03}	h_{04}	h_{05}
	0.0	0.0	0.5625	0.15	0.15	0.150
h_{0j}	0.0	0.25	0.25	0.25	0.51	0.050
	0.0	0.0	0.1875	0.6	0.34	0.80
	1.0	0.75	0.0	0.0	0.0	0.0
	h_{10}	h_{11}	h_{12}	h_{13}	h_{14}	h_{15}
	4.0	4.0	-1.25	0.0	0.0	0.0
h_{1j}	3.5	0.0	0.0	0.0	- 0.4	0.0
	3.0	3.0	1.25	0.0	0.4	0.0
	-10.5	-7.0	0.0	0.0	0.0	0.0
α_{0j}	1.0100	1.0275	1.0669	1.0875	1.0823	1.0915
α_{1j}	0.6750	0.4300	0.0625	0	0.0080	0
β_{0j}	0	0.0013	0.0055	0.0123	0.0089	0.0195
β_{2j}	0.6750	0.4300	0.0625	0	0.0080	0

The efficient frontier for the first parametric interval $0 \leq t \leq t_1$, appears to be parabolic. However, this is not the case and the efficient frontier is linear in mean-standard deviation space. To see why this is so, observe that $\beta_{00} = 0$ so that $\sigma_p^2 = \beta_{20} t^2$ or, $\sigma_p = \beta_{20}^{\frac{1}{2}} t$.

TABLE 9.4　　　Corner Portfolios for Example 9.3

j	0	1	2	3	4	5
t_j	0	0.0714	0.1071	0.33	0.6500	1.15
K_j	$\{6\}$	$\{2,6\}$	$\{5,2,6\}$	$\{4,5,2,6\}$	$\{4,5,6\}$	$\{3,4,5,6\}$
$\mu_p(t_j)$	1.0100	1.0582	1.0736	1.0875	1.0875	1.0915
$\sigma_p^2(t_j)$	0.0	0.0034	0.0062	0.0123	0.0123	0.0195
	$\hat{x}_0$	$\hat{x}_1$	$\hat{x}_2$	$\hat{x}_3$	$\hat{x}_4$	$\hat{x}_5$
	0.0	0.2857	0.4286	0.15	0.15	0.15
	0.0	0.25	0.25	0.25	0.25	0.05
$\hat{x}_j$	0.0	0.2143	0.3214	0.6	0.6	0.80
	1.0	0.25	0.0	0.0	0.0	0.0

With $\mu_p = \alpha_{00} + \alpha_{10}t$, this implies the efficient frontier is

$$\mu_p = \alpha_{00} + \beta_{20}^{\frac{1}{2}}\sigma_p,$$

in this first interval. This relationship between μ_p and σ_p is indeed linear in mean-standard deviation space and is similar to the CML. However, at the end of this linear relationship ($t = t_1 = 0.0714$), the holdings in the risk free asset have been reduced to 0.25 and are not reduced to zero until the end of the next interval ($t = t_2 = 0.1071$).

We complete this section by generalizing the results of the kink example (Example 9.1) to the general problem of (9.24). From Theorem 7.1, the efficient frontier for (9.24) is characterized by a finite number of intervals $[t_j,\ t_{j+1}]$, with $t_0 = 0$ and $t_j < t_{j+1},\ j = 0, 1, \ldots, N - 1$. In each parametric interval, $t_j \leq t \leq t_{j+1},\ j = 0, 1, \ldots, N - 1$, the efficient portfolios are

$$h_{0j} + th_{1j},$$

where h_{0j} and h_{1j} are obtained from the solution of the linear equations (7.31) which we repeat here

$$H_j \begin{bmatrix} h_{0j} \\ g_{0j} \end{bmatrix} = \begin{bmatrix} 0 \\ b_j \end{bmatrix} \quad \text{and} \quad H_j \begin{bmatrix} h_{1j} \\ g_{1j} \end{bmatrix} = \begin{bmatrix} \mu \\ 0 \end{bmatrix}. \quad (9.33)$$

In (9.33), $g_{0j} + tg_{1j}$ are the multipliers for the active constraints and

$$H_j = \begin{bmatrix} \Sigma & A_j' \\ A_j & 0 \end{bmatrix}, \quad (9.34)$$

where the columns of A'_j are the gradients of those constraints active in the interval.

The portfolio's expected return, μ_{pj}, is

$$\mu_{pj} = \alpha_{0j} + t\alpha_{1j}, \tag{9.35}$$

where

$$\alpha_{0j} = \mu'h_{0j} \quad \text{and} \quad \alpha_{1j} = \mu'h_{1j}.$$

Similarly, the portfolio's variance, σ^2_{pj}, is

$$\sigma^2_{pj} = \beta_{0j} + \beta_{1j}t + \beta_{2j}t^2, \tag{9.36}$$

where

$$\beta_{0j} = h'_{0j}\Sigma h_{0j}, \quad \beta_{1j} = 2h'_{0j}\Sigma h_{1j} \quad \text{and} \quad \beta_{2j} = h'_{1j}\Sigma h_{1j}.$$

These coefficients satisfy

$$\beta_{2j} = \alpha_{1j} \quad \text{and} \quad \beta_{1j} = 0. \tag{9.37}$$

Suppose for some interval $t_j \le t \le t_{j+1}$ that $\beta_{2j} = 0$. But then from (9.37), $\alpha_{1j} = 0$. From (9.35), (9.36) and (9.37), this means that both μ_{pj} and σ_{pj} remain constant throughout this interval. Furthermore, because $\beta_{2j} = h'_{1j}\Sigma h_{1j} = 0$ and Σ is positive definite, it follows that

$$h_{1j} = 0. \tag{9.38}$$

Expanding the defining equations (9.33) and (9.34) for h_{1j} gives

$$\Sigma h_{1j} + A'_j g_{1j} = \mu.$$

With $h_{1j} = 0$, this becomes

$$A'_j g_{1j} = \mu. \tag{9.39}$$

This last equation says there must *exist* a vector g_{1j} satisfying the equation. The number of equations in (9.39) is always n. However, the number of variables; i.e., the dimension of g_{1j} can vary between 1 and n. Examples of this dimension being just 1 are formulated in Exercise 9.3.

Our arguments so far are reversible. We summarize them in the following theorem.

Theorem 9.6 *Let $[t_j, t_{j+1}]$ be the j-th parametric interval defined by Theorem 7.1 for (9.24). The efficient frontier for (9.24) has a kink (1) if and only if $\beta_{2j} = 0$, and, (2) if and only if there exists a vector g_{1j} satisfying $A'_j g_{1j} = \mu$, where A_j is as described above.*

There is one circumstance in which it is easy to verify that the condition $A'_j g_{1j} = \mu$ of Theorem 9.6 is satisfied. That is when A_j is square. From Theorem 8.1(e), A_j has full row rank and so in this case, it is nonsingular. Thus it follows from (9.33) and (9.34) that

$$A_j h_{0j} = b_j \text{ and } A_j h_{1j} = 0,$$

which implies

$$h_{0j} = A_j^{-1} b_j \text{ and } h_{1j} = 0. \tag{9.40}$$

In this case, the efficient portfolios are $x(t) = h_{0j}$ for all t with $t_j \leq t \leq t_{j+1}$; i.e., $x(t)$ is constant for all t in this interval. Furthermore, by Definition 1.4, $x(t)$ is an extreme point for all t in this interval. Continuing the discussion of $x(t) = h_{0j}$ being an extreme point in the interval $[t_j, t_{j+1}]$, Theorem 9.2 asserts that $x(t)$ gives the maximum Sharpe ratio for (9.7) for all r_m satisfying

$$r_m = \mu' h_{0j} - \frac{h'_{0j} \Sigma h_{0j}}{t},$$

and t in the interval $[t_j, t_{j+1}]$. The smallest of these risk free returns corresponds to t_j and is

$$r_{m1} = \mu' h_{0j} - \frac{h'_{0j} \Sigma h_{0j}}{t_j}. \tag{9.41}$$

The largest of these risk free returns is

$$r_{m2} = \mu' h_{0j} - \frac{h'_{0j} \Sigma h_{0j}}{t_{j+1}}. \tag{9.42}$$

We conclude that $x(t) = h_{0j}$ gives the maximum Sharpe ratio for (9.7) for all r_m in the interval

$$r_{m1} \leq r_m \leq r_{m2}. \tag{9.43}$$

Similarly, we can apply Theorem 9.4. Again, let $x(t) = h_{0j}$ be an extreme point in the interval $[t_j, \ t_{j+1}]$. Theorem 9.4 states that $x(t)$ gives the maximum implied risk free return when the Sharpe ratio, R, given by

$$R = \left[\frac{(h'_{0j} \Sigma h_{0j})^{\frac{1}{2}}}{t} \right],$$

and t is in the interval $[t_j, \ t_{j+1}]$. The smallest of these Sharpe ratios corresponds to t_{j+1} and is

$$R_1 = \left[\frac{(h'_{0j} \Sigma h_{0j})^{\frac{1}{2}}}{t_{j+1}} \right]. \tag{9.44}$$

The largest of these is

$$R_2 = \left[\frac{(h'_{0j} \Sigma h_{0j})^{\frac{1}{2}}}{t_j} \right]. \tag{9.45}$$

We conclude that $x(t) = h_{0j}$ gives the maximum implied risk free return when the Sharpe ratio R is in the interval

$$R_1 \leq R \leq R_2. \tag{9.46}$$

We summarize these results in the following theorem.

Theorem 9.7 *Let $[t_j, \ t_{j+1}]$ be the j-th parametric interval defined by Theorem 7.1 for (9.24). Assume the efficient portfolio for this interval is an extreme point for (9.24) so that $x(t) = h_{0j}$ is an efficient portfolio for all $t \in [t_j, \ t_{j+1}]$. Then*

(a) $x(t) = h_{0j}$ gives the maximum Sharpe ratio for (9.7) for all r_m in the interval $r_{m1} \leq r_m \leq r_{m2}$, where r_{m1} and r_{m2} are defined by (9.41) and (9.42), respectively.

(b) $x(t) = h_{0j}$ gives the maximum implied risk free return when the Sharpe ratio R is in the interval $R_1 \leq R \leq R_2$, where R_1 and R_2 are defined by (9.44) and (9.45), respectively.

9.3   Computer Results

Figure 9.2 shows the MATLAB code which uses PQPSolver to solve
Example 9.2. Other than lines 3 to 8 which define the data, this program
is identical to Example8p1 shown in Figure 8.1. The reader can see the
discussion of this program in Section 8.2 prior to Figure 8.1.

```
1   %Example9p2.m
2   % 3 asset problem which has a kink in the efficient
3   frontier
4   A = [   1 0 0; 0 1 0; 0 0 1; 0 1 1; 1 1 1 ]
5   b = [ 0.8 0.25 0.8 0.85 1 ]'
6   mu = [ 1.05 1.08 1.1 ]'
7   Sigma = [1 0 0; 0 2 0; 0 0 3]/100.
8   n = 3; m =   4; q = 1;
9   x0 = [0.2 0.2 0.6 ]';
10  checkdata(Sigma,1.e-6);
11  checkconstraints(A,b,x0,m,q,n);
12  [H0j,H1j,U0j,U1j,tvec,numint] = ...
13                      PQPSolver(A,b,m,n,q,x0,Sigma,mu);
14  msg = 'Efficient Frontier Completely Determined'
15  tvec
16  H0j
17  H1j
18  U0j
19  U1j
20  Alpha0 = []; Alpha1 = []; Beta0 = []; Beta2 = [];
21  for  j = 1:numint
22      Alpha0(j) = mu' * H0j(1:n,j);
23      Alpha1(j) = mu' * H1j(1:n,j);
24      Beta0(j)  = H0j(1:n,j)' * Sigma * H0j(1:n,j);
25      Beta2(j)  = H1j(1:n,j)' * Sigma * H1j(1:n,j);
26  end
27  Alpha0
28  Alpha1
29  Beta0
30  Beta2
31  %  Expected return and variance at start of each interval
32  for  j = 1:numint
33      mu(j)  =  Alpha0(j) + tvec(j)*Alpha1(j);
34      sig2(j) = Beta0(j) + tvec(j)*tvec(j)*Beta2(j);
35  end
36  mu
37  sig2
```

Figure 9.4.   Example 9.2: Example9p2.m.

```
1   %Example9p3.m
2   % 4 asset problem including a risk free asset
3   % based on a 3 asset problem which has a kink
4   % in the efficient frontier
5   A = [  1 0 0 0; 0 1 0 0; 0 0 1 0; 0 1 1 0; 0 0 0 -1;
6           1 1 1 1 ]
7   b = [ 0.8 0.25 .8 0.85 0 1 ]'
8   mu = [ 1.05 1.08 1.1 1.01 ]'
9   Sigma = [1 0 0 0; 0 2 0 0; 0 0 3 0; 0 0 0 0]/100.
10  n = 4; m = 5;   q = 1;
11  x0 = [0.2 0.2 0.6 0.0]';
12  %checkdata(Sigma,1.e-6);
13  checkconstraints(A,b,x0,m,q,n);
14  [H0j,H1j,U0j,U1j,tvec,numint] = ...
15                      PQPSolver(A,b,m,n,q,x0,Sigma,mu);
16  msg = 'Efficient Frontier Completely Determined'
17  tvec
18
19  H0j
20  H1j
21  U0j
22  U1j
23  Alpha0 = []; Alpha1 = []; Beta0 = []; Beta2 = [];
24  for j = 1:numint
25      Alpha0(j) = mu' * H0j(1:n,j);
26      Alpha1(j) = mu' * H1j(1:n,j);
27      Beta0(j)  = H0j(1:n,j)' * Sigma * H0j(1:n,j);
28      Beta2(j)  = H1j(1:n,j)' * Sigma * H1j(1:n,j);
29  end
30  Alpha0
31  Alpha1
32  Beta0
33  Beta2
34  %  Expected return and variance at start of each interval
35  for j = 1:numint
36     mu(j) =  Alpha0(j) + tvec(j)*Alpha1(j);
37     sig2(j) = Beta0(j) + tvec(j)*tvec(j)*Beta2(j);
38  end
39  mu
40  sig2
```

Figure 9.5. Example9p3.m, Risk Free Return $r = 1.01$.

9.4   Exercises

9.1 Let (σ_m, μ_m) be a point on the efficient frontier for (9.24) which is not a kink point.

(a) Show that the implied risk free return r_m is

$$r_m = \mu_m - \frac{\alpha_1 \sigma_m^2}{[\alpha_1(\sigma_m^2 - \beta_0)]^{\frac{1}{2}}}.$$

(b) In Example 9.1, the efficient frontier does have a left-hand derivative for $t = t_1$ and a right-hand derivative for $t = t_2$. Use the coefficients in Table 9.1 and Theorem 4.1(a) to show they are identical to r_{m1} in (9.28), and r_{m2} in (9.29).

9.2 Let Σ be positive semidefinite and let A be (m, n). Assume A has full row rank. Let

$$H = \begin{bmatrix} \Sigma & A' \\ A & 0 \end{bmatrix}.$$

(a) Prove that H is nonsingular if and only if $s'\Sigma s > 0$ for all $s \neq 0$ with $As = 0$.

(b) Show that if Σ is positive definite then any matrix of the form

$$H = \begin{bmatrix} \begin{bmatrix} \Sigma & 0 \\ 0' & 0 \end{bmatrix} & A' \\ A & 0 \end{bmatrix},$$

where A is $(m, n+1)$, is nonsingular provided A has a row containing $(l', 1)$ and l is an n-vector of ones.

9.3 Apply Theorem 9.6 to each of the problems (2.2) and (2.3) under the assumption that there is a constant θ such that $\mu = \theta l$. Explain your results in light of your solution for Exercise 2.9.

9.4 Use Table 9.1 to verify the results of Theorem 8.2 applied to Example 9.2.

9.5 Use Table 9.3 to verify the results of Theorem 8.2 applied to Example 9.3.

Appendix

In this appendix, we summarize the name, purpose, and page reference for each of the computer programs presented throughout the text. The names of the various routines are formulated so that "EF" refers to the Efficient Frontier, "MV" refers to Mean Variance, "MSD" refers to Mean Standard Deviation and "CML" refers to Capital Market Line.

TABLE 10.1   **Summary of Computer Programs**

Chap	Name	Purpose	Page
1	Example1p2.m	Solves $\min\{c'x + \frac{1}{2}x'Cx \mid Ax = b\}$	17
	checkdata.m	checks for C being symmetric and positive definite	18
2	Example2p1.m	Calls EfficFrontCoeff to obtain efficient frontier coefficients for Example 2.1	34
	EFMVcoeff.m	Calculates coefficients of efficient frontier in Mean Variance space	34
	EFMVplot.m	Plots the efficient frontier using the coefficients from EFMVcoeff.m	33
3	Example3p1.m	Calls CMLplot to plot the CML for Example 3.1	54
	CMLplot.m	Plots market portfolio and CML in MSD space using EFMSDplot.	56
	EFMSDplot.m	Plots efficient frontier in Mean-Standard Deviation Space.	57
4	SharpeRatios.m	Calculates Sharpe ratios and associated risk free rates.	75

TABLE 10.1   **Summary of Computer Programs** *Continued*

References

1. Arseneau, Lise and Best, Michael J., "Resolution of degenerate critical parameter values in parametric quadratic programming," Department of Combinatorics and Optimization Research Report CORR 99−47, University of Waterloo, Canada, 1999.

2. Best, Michael J., "Equivalence of some quadratic programming algorithms," *Mathematical Programming* **30**, 1984, 71−87.

3. Best, Michael J., and Hlouskova, Jaroslava, "Quadratic programming with transaction costs," *Computers and Operations Research* (Special Issue: Applications of OR in Finance), Vol 35, No. 1, 18−33, 2008.

4. Best, Michael J., and Hlouskova, Jaroslava, "An Algorithm for Portfolio Optimization with Variable Transaction Costs I: Theory," *Journal of Optimization Theory and Applications*, Vol. 135, No. 3, 2007, 563−581.

5. Best, Michael J., and Hlouskova, Jaroslava, "An Algorithm for Portfolio Optimization with Variable Transaction Costs II: Computational Analysis," *Journal of Optimization Theory and Applications*, Vol. 135, No. 3, 2007, 531−547.

6. Best, Michael J., and Hlouskova, Jaroslava, "Portfolio selection and transactions costs," *Computational Optimization and Applications*, 24 (2003) 1, 95−116.

7. Best, Michael J., and Hlouskova, Jaroslava, "The efficient frontier for bounded assets," *Mathematical Methods of Operations Research*, 52 (2000) 2, 195−212.

8. Best, Michael J., "An algorithm for the solution of the parametric quadratic programming algorithm," in *Applied Mathematics and Parallel Computing − Festschrift for Klaus Ritter*, H. Fischer, B. Riedmüller and S. Schäffler (editors), Heidelburg: Physica−Verlag, 1996, 57−76.

9. Best, Michael J., and Hlouskova, Jaroslava, "An algorithm for portfolio optimization with transaction costs," *Management Science*, Vol 51, No. 11, November 2005, 1676−1688.

10. Best, Michael J. and Grauer, Robert R., "Positively weighted minimum-variance portfolios and the structure of asset expected returns," *Journal of Financial and Quantitative Analysis*, Vol. 27, No. 4 (1992) 513−537.

11. Best, Michael J., and Grauer, Robert R., "On the sensitivity of mean-variance-efficient portfolios to changes in asset means: some analytical and computational results," *The Review of Financial Studies*, Vol. 4, No. 2 (1991) 315−342.

12. Best, Michael J., and Grauer, Robert R., "Sensitivity analysis for mean-variance portfolio problems," *Management Science*, Vol. 37, No. 8 (1991) 980−989.

13. Best, Michael J. and Grauer, Robert R., "The efficient set mathematics when the mean variance problem is subject to general linear constraints," *Journal of Economics and Business*, 42 (1990) 105−120.

14. Best, Michael J. and Grauer, Robert R., "Capital asset pricing compatible with observed market weights," *Journal of Finance*, Vol. XL, No. 1 (1985) 85−103.

15. Best, Michael J., and Grauer, Robert R., "The analytics of sensitivity analysis for mean variance portfolio problems," *International Review of Financial Analysis*, Vol. 1, No. 1 (1991) 17−37.

16. Best, Michael J. and Kale, J., "Quadratic programming for large-scale portfolio optimization," in *Financial Services Information Systems*. (2000)

17. Best, Michael J., and Ritter, Klaus, *Linear Programming: Active Set Analysis and Computer Programs*, Prentice-Hall, Inc., Englewood Cliffs, New Jersey, 07362.

18. Mangasarian, O. L., *Nonlinear Programming*, McGraw-Hill, New York, 1969.

19. Markowitz, Harry M., *Portfolio Selection: Efficient Diversification of Investments*, Coyles Foundation Monograph, Yale University Press, New Haven and London, 1959. This has been more recently reprinted as, Markowitz, Harry M., *Portfolio Selection*, Blackwell Publishers Inc., Oxford, UK, 1991.

20. Noble, Ben, *Applied Linear Algebra*, Prentice-Hall, Inc., Englewood Cliffs, New Jersey, 1969.

21. Perold, Andre F., "Large-scale portfolio optimization," *Management Science* **30** No. 10 1984 1143−1160.

22. Sharpe, William F., *Portfolio Theory and Capital Markets*, McGraw-Hill, 1970.

23. Sherman, J. and Morrison, W. J., "Adjustment of an inverse matric corresponding to changes in the elements of a given column or a given row of the original matrix," *Ann. of Mathematical Statistics*, **20** 1949 621.

24. van de Panne, C. and Whinston, A., "The symmetric formulation of the simplex method for quadratic programming," *Econometrica* **37** 1969 507−527.

25. Woodbury, M., "Inverting modified matrices," Memorandum Report 42, Statistical Research Group, Princeton University, New Jersey (1950).

26. Wang, Xianzhi, "Resolution of Ties in Parametric Quadratic Programming," Masters Thesis, Department of Combinatorics and Optimization, University of Waterloo, Waterloo, Ontario, Canada, 2004.

Index

221